Blockchain Business Strategy

With Case Studies for the Real World Economy

Blockchain Business Strategy

A Guide for Managers to Disentangle and Apply Blockchain Technology

Michael F. Mazier

A Silent Killdeer Publishing (asilentkilldeer.com)

New York • New Orleans • Milan • Moscow

Blockchain Business Strategy

Written by Michael F. Mazier

Edited by Russell J.T. Dyer, Shula Klinger.

A Silent Killdeer Publishing

Published by A Silent Killdeer Publishing, New Orleans, Louisiana, U.S.A. Visit our web site at http://asilentkilldeer.com or email us at inquiries@asilentkilldeer.com. Printed in the United States of America. Set in Georgia font.

ISBN-13: 978-0-9831854-8-2

Thanks to my parents, Virgilio and Virginia Mazier, whose sacrifices, resilience, and unwavering faith have instilled within me a profound appreciation for the precious gift of liberty and search for knowledge.

Table of Contents

Preface 1

Introduction 3

Part One, Decentralization 9

- Ancient Decentralization 13
- Modern Decentralization 23

Part Two, Business Strategy 33

- Market Power 39
- Barriers to Entry 47

Part Three, Case Studies 57

- Case 1: Digital Jukebox 61
- Case 2: Trading Bills of Lading 79
- Case 3: Car-Sharing Service 89
- Case 4: Real Estate Finance 99
- Case 5: Monetizing Private Data 109
- Case 6: Gold-Backed Token 121

Conclusion 141

Preface

Despite introducing many theories and concepts, this book has a practical purpose. It demonstrates, in a step-by-step fashion, how strategic thinking can be used to devise blockchain applications.

We'll start by referring to the original Bitcoin white paper to define blockchain technology. The Bitcoin creator presented it as a technology that *allows any two willing parties to transact directly with each other without the need for a trusted third party.*[1]

Many books hype blockchain as a revolutionary, democratizing technology that will change the world, but they don't show us how this revolution might come about. They're filled with lofty expressions like "new paradigm," "unleashing the potential," and "the power to transform." Such promises and predictions are tantalizing, but they generally lack substance.

I wrote this book to bridge the gap between the highly technical and the highly promising. It responds to practical questions and comments from friends and associates, such as: "How does blockchain help in my business?" "We tried a blockchain pilot project and it failed." It is neither a book about trading bitcoin nor programming. Instead, it's about doing the right thing before doing things right.

Bitcoin has become increasingly popular in the financial economy, with investors able to buy, sell, and trade bitcoin through

[1] Nakamoto, Satoshi. "Bitcoin: A Peer-To-Peer Electronic Cash System." 31 Oct 2008.

exchange-traded funds or other financial instruments. However, blockchain technology is yet to be widely adopted in the real economy, which involves producing and consuming goods and services. The financial and real economies are interconnected, and the growth of one affects the other.

In the traditional financial economy, profits are generated via centralized control of transactions under the scrutiny of government regulators, who compel centralization. However, blockchain technology is designed to eliminate the need for trusted third parties. Blockchain applications in the real economy will almost surely follow the spirit of Bitcoin - disempowering intermediaries. Thus, blockchain-related activities in the financial and real economies have contrary propensities.

To determine how to apply the capabilities of blockchain technology, we must unbundle, disentangle, and re-engineer existing business models — or create entirely new business models that were not previously possible.

Introduction

Since prehistoric times, our ancestors have been socializing and trading face-to-face. The arrival of the internet helped us create a whole world of simulated interactions in which we spend an inordinate amount of time each day gazing at electronic devices. This technology helped create simulated banking, in which software and devices replace face-to-face money transactions. Now, a financial institution interposes itself between us. They've become ubiquitous third-party intermediaries.

We've also come to rely on social media corporations as intermediaries to simulate face-to-face socializing. It's not actual socializing, though. However, it serves the same purpose — to some extent. In exchange for free services, social media corporations gather and sell privileged knowledge about our day-to-day lives, personal habits, political views, and even our deepest personal thoughts — often without us realizing it.

Other ramifications go beyond abstract financial or technical notions. They involve the inevitable human impact of changes to the economic and technological landscape, particularly shifts in power relationships in politics and business.

Individuals and entities with more power can exert their will over the less powerful or the unaware (i.e., the general public). They do this not with brute force but quietly and subtly by controlling access to technology, knowledge, or money.

In the business realm, the more powerful entities have market power: the ability to raise prices above the level that would prevail under competitive circumstances. Market power measures a firm's control or dominance over its market.

Third-party intermediaries are in ideal positions to amass secret or privileged knowledge and gain a high degree of market power. Consider that a bank customer can be charged fifty dollars for a wire transfer of funds[1] even though the customer does all the work (i.e., go online and complete the instructions). No bank employee is involved. This suggests that some third-party intermediaries earn fees above the costs of providing their services. Credit reporting agencies are also data brokers. Equifax, for example, advertises data related to individuals' automobile use, restaurant activity, insurance plans, travel and leisure preferences, and more.[2] Intermediaries can sell privileged information about our private transactions.

The creator of Bitcoin, Satoshi Nakamoto, introduced a form of electronic cash that allowed payments to be sent directly from one party to another without going through a third-party financial institution.

Embedded in the computer code of the first bitcoin transaction on January 3, 2009, was the day's headline in the Financial Times of London: "Chancellor on Brink of Second Bailout for Banks" - an outcry by Satoshi against combined political and market power. The original Bitcoin paper introduced a new wave of thinking about finance and technology that cleared the way for disintermediation: circumventing powerful intermediaries, the antiquated middleman.

Blockchain technology does indeed facilitate the development of new business models that previously were not possible. At the

[1] Goldberg, Matthew. "How Much Are Wire Transfer Fees?" Bankrate, 30 Nov 2023.
[2] Sherman, Justin. "Credit Reporting Agencies Don't Just Report Credit Scores." Duke University, Sanford School of Pubic Policy, Tech Policy Lab, 9 Nov 2022.

same time, blockchain technology destroys the market power of third-party intermediaries and distributes knowledge and power across the economy to end-users. Nevertheless, someone still needs to perform transactional tasks done by third-party intermediaries — albeit at competitive prices. Somebody still needs incentives to build, run, and maintain a blockchain network. Blockchains are designed to operate on a competitive system of rewards and incentives, with transaction fees that should cover the costs of running the network — but not much more.

Don't think, though, that blockchain technology is a panacea. It will not make banks obsolete. Instead, it will change what they do. Banks and other third-party intermediaries perform a bundle of services. Blockchain technology can only be applied to some of these services.

We need to examine what it means to execute a transaction, to make purchases, and to be paid for them. We need to consider how funds travel, what constitutes currency, and who is interposing, surveilling, or taking a cut as we transact — particularly online.

For example, consider how Amazon built its business model in the 1990s. It was entirely online, with no brick-and-mortar stores. As a result, Amazon has a positive cashflow before sending out the goods. By contrast, brick-and-mortar stores prefunded their inventory and waited weeks after stocking and turning over their inventory for positive cash flow[3]. Just as Amazon used the internet as an integral part of its business model, blockchain technology needs to be woven seamlessly into the fabric of a business — not added as an afterthought.

[3] Fox, Justin. "At Amazon, It's All about Cash Flow." Harvard Business Review, 20 Oct. 2014.

Organization of Book

This book has three parts: an overview of decentralization, business strategy, and six case studies.

Decentralization: This overview of ancient and modern decentralization starts with a story about decentralized money exchange. Centuries ago, Pacific islanders on Yap created a monetary system where two parties could transact without a third-party intermediary. Today, we can see advances in cryptography and networking that allow two parties to conduct transactions with no third-party intermediary, thanks to blockchain technology.

This is critical because, between the development of the Yap economy and today, we have relied heavily on banks and other institutions to store, move, and track our wealth. When decentralized exchanges occur, these third-party intermediaries lose their influence, control, and power over buyers and sellers. It is the nature of blockchain to destroy the market power wielded by centralized third-party intermediaries to the benefit of end-users.

Strategy: This business strategy overview includes commentary on how it can apply to blockchain technology. It explains market power and explores trends in how companies have gained and maintained dominance since the 1980s. Then, we focus on competitive advantages: what they are, and how to achieve these elusive positions. We'll look at how barriers to entry restrict competition. We'll discuss how to secure a situation and method that leads to long-term superior performance. To do this, we'll explore six types of competitive advantages.

Case Studies: This last part presents six case studies, each from the point of view of a person trying to solve a practical, real-world problem. Each case focuses on one of the competitive advantages defined in the previous part and can be read on a stand-alone basis.

Hopefully, this book inspires brainstorming sessions that focus on the business-oriented characteristics of blockchain technology, sparking your creativity at the crossroads of business and technology. We follow the adage of do the right thing before you do things right. Before we devise a blockchain application, we look at the big picture of market power, competitive advantages, and *barriers to entry* among competing firms. How do they acquire and use market power? What barriers to entry are there, and which ones should we seek to use in devising a winning business model? You'll understand why decentralization is an opportunity or threat from a business perspective for new or legacy companies.

This book covers primarily technology and economics, but it's ultimately about business strategy. It's an interdisciplinary work that should help you better understand blockchain, clarify your goals, and design plans with sound judgment to prepare for a world with blockchain technology.

PART ONE

With most people nowadays connected to electronic devices, our daily experiences depend on a central authority, such as a bank (see Figure 1). The general population — let's call them Alice and Bob — can transact with each other. However, behind the scenes, their transactions pass ubiquitously through a central authority that collects transaction fees, harvests our private information, and has the power to censor transactions.

Figure 1

FROM CENTRALIZATION TO DECENTRALIZATION

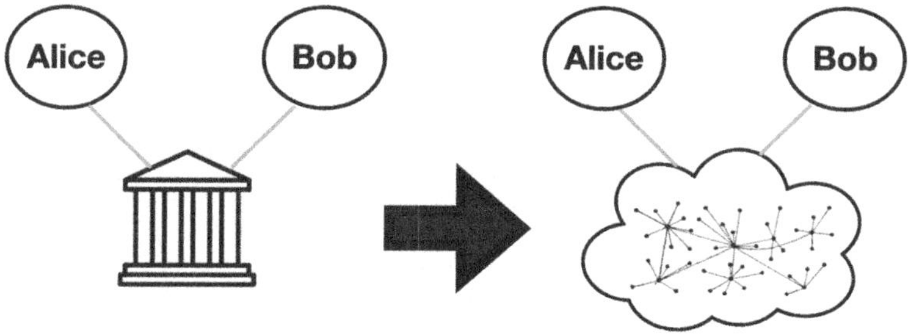

Central authorities may charge excessive fees or economic rents, which is economic-speak for a price that exceeds the cost of delivering their goods or services. Perfect competition keeps fees at cost. A central authority without competition has no incentive to lower fees. Alice and Bob might complain about excessive fees. They may never read the fine print agreement about how their private data will be shared and resold. They might switch banks, but everything will mostly stay the same. Still, they're happy that they're taking care of business.

Now consider the right side of Figure 1. Here, Alice and Bob can transact with each other. Behind the scenes, their transactions pass through a competing crowd of verifiers — called miners in Bitcoin — forming a group consensus, each competing to win transaction fees. Transaction fees resulting from stiff competition are closer to the ideal of perfect competition. Transaction fees should cover the cost of running the network and not much more. The verifiers still need to be paid, but none has the market power to charge fees much higher than costs because a competing verifier would step in to process the transaction under a highly competitive market structure. Competition keeps prices low. Hence, a blockchain solution destroys the market power of a central authority, benefiting end-users Alice and Bob.

The chapters of Part I provide an overview of decentralization, where transactions are verified and recorded by a network of participants rather than a central authority.

Chapter 1 describes the story of an ancient form of decentralization on the island of Yap in the Pacific. Many centuries ago, they used stones as currency, each weighing over a ton. The gigantic stones were stationary, like gold held in a vault. They traded value via word of mouth without the need for a central authority. The underlying principles of Yap's system are analogous to modern-day blockchain.

Chapter 2 discusses implementing a Yap-like system using a network of computers without central control. On Yap, everyone knew each other, and transactions were conducted face-to-face. However, people can fake their identities or create bogus transactions on the faceless internet. The problem of transferring electronic cash without third-party intermediaries was so difficult that it took until 2008 for Bitcoin to be created.

Chapter One

Ancient Decentralization

Transferring value without a central authority isn't new. Centuries ago, on the Yap island in the Pacific, about five-hundred miles south of Guam, a monetary system was used with a decentralized exchange of value.

Figure 2

Big Money

Drawing[1] of Yapese Stone Money

[1] "Drawing from Hamburger Südsee-Expedition 1908-1910" Wikimedia Commons, Wikimedia Foundation.

The Yapese traded large circular carved limestones called *Rai* stones, each weighing up to four tons. Rai stones represented money. These gigantic coins remained stationary and were displayed around the island. The Yapese all knew who owned each stone; transactions of stones were by word of mouth.

Since it was very labor-intensive to source new Rai stones, there was little chance of currency dilution. Limestones were not native to Yap, so the workers had to sail hundreds of miles to the nearby islands of Guam or Palau to quarry and carve limestones. They then lugged the tons of stones back to Yap on boats. The stones were valuable because everyone in the community knew how much hard work was needed to bring stones to Yap. Compare this form of currency to other cultures that might use seashells found on the beach.

According to legend, one particular expedition encountered stormy seas, and the sailors were forced to cut a stone loose if they were to stay afloat. The Rai stone was lost at the bottom of the Pacific Ocean forever, but the sailors were able to struggle back home to Yap and share their harrowing experiences. The validity of the sailors' story was accepted on Yap. Despite the stone being physically absent, the community still recognized its ownership, and it became part of the monetary base while resting at the bottom of the Pacific Ocean.

Rai stones seem impractical to outsiders who think of money as a physical object that one could carry in a pocket. However, like digital money today, Rai stones underpin a system in which participants own something of intrinsic value without holding a physical object. In a sense, the money is virtual. If owners of a Rai stone decided to transfer value to someone else, they would announce the transfer to the entire community. Everyone would hear of this and collectively acknowledge that a particular Rai stone had a new owner. The whole community participated in transferring ownership to the latest recipient. The transfer was validated because the community members shared a common understanding.

In economic terms, Yap mitigated the risk of inflation because it was so difficult to increase the money supply. That is why the Rai stones were a highly effective store of value. The value of each Rai stone, like the value of a diamond, was based on size, artistry, the beauty of the stone, or its history. Every member of the Yap community made a mental ledger of ownership for each unique Rai stone, with each stone's history contributing to its value.

Everyone knew if a particular Rai stone had once been gifted as a wedding dowry or owned by a prominent family. This knowledge helped everyone to remember the unique features of each stone. When ownership of a particular Rai stone was transferred, the entire community would update their mental ledgers with the new owner of the stone, updating their mental history of the stone. Again, a transfer of ownership occurred without a central authority.

This ancient system set a precedent for transferring monetary value without a central authority or a bank. In the 1991 paper "The Island of Stone Money,"[2] Nobel laureate economist Milton Friedman wrote how Rai stones for the Yapese were equivalent to the gold standard, the monetary system favored by Friedman, in which the value of a currency is based on a fixed quantity of gold kept in reserve.

In short, ancient cultures around the world invented money independently as a step beyond the barter system. At various times and places, money has taken the form of seashells, stones,

[2] Friedman, Milton."The Island Of Stone Money," Working Papers in Economics, no. E-91-3. Stanford, California: Hoover Institution, 1991.

ropes, precious metals, paper, or digital entries in a computer system. On their isolated island that lacked gold or silver, the islanders of Yap were unique in using multi-ton objects to represent money as a medium of exchange and a store of value.

In 1954, the actor Burt Lancaster starred in "His Majesty O'-Keefe," a movie about Yap. Based on a true story, the film shows a shipwrecked sailor arriving on Yap in 1871. Having learned about the Rai stone economy, he helps locals produce more stones with machinery. He then trades them for dried coconut kernels to make coconut oil. For the islanders, these newer stones are worth less than the older stones produced by traditional, labor-intensive methods. Regardless, the new stones contribute to an increase in the money supply, resulting in inflation. For the economy to recover, the new stones must be destroyed. Or, in macroeconomic terms, an end to erratic growth of the money supply reduced hyper-inflationary fears—confidence in money returned with steady, predictable growth in the money supply.

Although fictional, the movie demonstrates three critical points: First, the meaning of *value* is communally and culturally determined. On Yap, value was determined by human labor—hence the drop in the value of new, machine-made stones. Second, trust and confidence are critical factors in any monetary system, distributed or otherwise. Whether the currency is gold, bank notes, or giant stones doesn't matter. And third, a replicated ledger of ownership lives in the minds of the community.

Why Stone Money Matters

While the monetary system of Yap was established hundreds of years ago, it offers valuable lessons for students of today's economy. Its nature as a decentralized system of exchange functions as a model for transactions conducted via blockchain. In this way, a few relevant observations can be made:

▮ OBSERVATION 1: Money is a social construct.

Money exists and has value because people agree that this is so. Money exists strictly in the minds of people in society — however far apart they're located.

▮ OBSERVATION 2: No one gains political or market power.

No monarch or government has a monopoly on money by force of law with the potential to abuse political power. No one can print or devalue money — a potential form of taxation without representation. No centralized financial intermediaries can charge transaction fees above the level that would prevail under competitive circumstances. No one has the capability to exploit private information or gain unfair advantages in the market.

▮ OBSERVATION 3: Yap was a cashless society.

No one carried cash on their person. The gigantic Rai Stones were analogous to gold held in reserve. However, no central banker maintained a record of ownership. Instead, every community member memorized a record of ownership of all Rai stones. In accounting terms, these records were ledgers.

▮ OBSERVATION 4: Yap used a distributed ledger.

Every community member maintained a replicated mental ledger of the entire monetary system. Like the DNA replicated throughout all the cells in our bodies, an identical mental ledger of all transactions is replicated and distributed into the consciousness of all the islanders on Yap.

▮ OBSERVATION 5: Ledgers were updated via decentralized trust.

No single person is trusted to verify transactions and then dictate how the crowd must update their mental ledgers. Trust is placed in the crowd as a whole. The crowd must first agree by acclamation to verify a new transaction. Only then do they each update their own mental ledgers, ensuring that all ledgers remain synchronized with one another. The result is a single version of the truth stored in replicated ledgers in everyone's consciousness.

▮ OBSERVATION 6: No double-spending.

The same sender cannot spend the same Rai stone twice because the crowd wouldn't approve the second transaction. Double spending is easy to avoid with a trusted third-party intermediary controlling a centralized ledger. However, only with the recent invention of Bitcoin have we seen a way to prevent double-spending in the digital realm without a centralized third-party intermediary.

▮ OBSERVATION 7: Rai stones require proof of work.

Proof of work is an economic measure that demonstrates the expenditure of resources, which is manual labor in the case of Yap. Upon inspection of a stone, a Yapese could determine immediately that plenty of work had been expended in carving a single piece of limestone into a circularly shaped Rai stone with a hole in its center. The community would accept into its money supply only stones that evidenced significant expenditure of work. This difficulty in producing Rai stones limits the total money supply of stones, which in turn controls inflation.

The term proof of work was coined in a Bell Labs research paper in 1999,[3] based on an earlier work by Cynthia Dwork and Moni Naor[4]. Dwork and Naor described a method to deter spam email. The proposed email system required that senders attach to their emails proof of work: the solution to a moderately difficult computational puzzle, devised to expend processing time in a computer. The recipient accepts only emails with valid proof of work (i.e., the solution to the puzzle). A user sending a legitimate email would hardly notice the extra computer processing time to compute proof of work for a single email. However, a spammer sending thousands of junk emails would need to expend an arduous amount of computer processing time.

A key feature of proof of work is the asymmetry between doing the work and verifying that the work was performed. The asymmetry on Yap is that it was difficult to produce a stone, but easy to verify that plenty of work had been expended in producing a stone. In the Burt Lancaster film, when the outsider O'Keefe found a more efficient and quicker way to produce Rai stones, the community devalued his newer stones relative to the older, manually created stones. The O'Keefe stones had less proof of work.

▮ OBSERVATION 8: Money supply growth is highly predictable.

The number of workers on the island provided a practical limit on the number of stones that could be produced. O'Keefe's

[3] Jakobsson, M., Juels, A. (1999). Proofs of Work and Bread Pudding Protocols(Extended Abstract). In: Preneel, B. (eds) Secure Information Networks. IFIP — The International Federation for Information Processing, vol 23. Springer, Boston, MA.

[4] Dwork, C., Naor, M. (1993). Pricing via Processing or Combatting Junk Mail. In: Brickell, E.F. (eds) Advances in Cryptology — CRYPTO' 92. CRYPTO 1992. Lecture Notes in Computer Science, vol 740. Springer, Berlin, Heidelberg.

addition of stones to the money supply and its ensuing runaway inflation were short-lived before things returned to normal.

▮ OBSERVATION 9: No one needs permission to join the system.

No gatekeeper exists to grant permission, certification, or authorization for anyone to memorize their ledgers and verify transactions; the more, the merrier. There is no social mechanism to exclude a person from the network. The analogy to a blockchain system is that if someone had the power to serve as a gatekeeper, that person would become a central authority because they could appoint members to bias the results. A gatekeeper who grants permission can control a network like a dictator with puppet legislators can control a country.

▮ OBSERVATION 10: The system offers non-repudiation.

Transactions cannot be denied or undone. If an islander announced the transfer of his ownership rights, no one could deny the fact of it. The Yap community may ostracize someone who denied having transferred a stone.

▮ OBSERVATION 11: Miners perform a social good.

Miners who quarry and produce Rai stones incur an opportunity cost by not performing otherwise useful work on the island — or giving up leisure time. Yet the system evolved whereby the miners' contribution must have been rewarded in some way with proceeds from newly minted Rai stones. At the very least, they must have been paid subsistence wages or extra compensation for the risk of sailing hundreds of miles with multi-ton stones. Miners contributed a social good to society by helping to maintain a money system resistant to excessive political and economic power concentrations. In this system, a central banker would be redundant because the miners and the community of verifiers fulfill that role.

▮ OBSERVATION 12: The system is always on.

Oral tradition would pass on the mental distributed ledger from generation to generation. Institutionalized memory matters more than the memory of any single individual. If someone gets sick or is traveling away from the island, they are, in modern parlance, offline. Upon returning to everyday life, they would reconnect to the network by asking for any transactions they may have missed. Now, their mental ledgers are back in sync with the community. If someone needs to remember their mental ledger, they will go along with the majority view and refresh their memory with what the majority had verified. In a modern Bitcoin network sense, this would be like computers that go offline and then update their ledger upon reconnecting to the network.

▮ OBSERVATION 13: Transactions appended only.

Like adding new links to the end of a chain, each subsequent transaction only builds upon the prior history. The Yapese would have recalled the ownership history of each stone in a time-sequential order. Perhaps a son was bequeathed a stone from his father, a stone that had remained in their family for many generations. The latest owner of a stone is the valid one, superseding all prior owners of the same stone.

▮ OBSERVATION 14: No single point of failure.

No single person can break the system, intentionally or unintentionally. A single forgotten transaction would not affect the crowd's consensus. No one can sabotage, embezzle, or skim money. It would require a majority of the community to alter ownership of a Rai stone.

❦

We can't be sure if the creator of Bitcoin studied the Yap money system. Nevertheless, suppose we tried to digitize Yap's system of mental ledgers on the internet. The conventional way would involve a computer network and a centralized ledger controlled by a third-party intermediary, such as a bank. This traditional online banking approach would force Yap to abandon its decentralized way of transferring value. The introduction of Bitcoin was the breakthrough that enabled decentralized electronic ways of transferring value.

Chapter Two

Modern Decentralization

In 2008, the Bitcoin whitepaper introduced a breakthrough: a form of electronic cash that could operate without the need for a third-party intermediary. It did not rely on a government or financial institution. Not since the days of stone money on Yap has there been a decentralized way to transfer value without handling cash. Before Bitcoin, it was not possible to replicate a Yap-like system in software to create decentralized cash. Unlike the real-life Yap stone system where users face each other, the internet exposes users to anonymous hackers or bogus transactions, such as spending the same money twice.

To explore how blockchain works, we'll start with a naive software implementation of the Yap system and then address limitations until we reach the modern blockchain breakthrough.

Suppose we build a simple software version of the Yap system on a connected computer network. Since it's on the internet, we could publish our software code for anyone to join the network. Suppose that Alice wants to send cash to Bob. On Yap, she would announce her Rai Stone transaction to the crowd of islanders, who would collectively verify her transaction and transfer ownership of a stone from Alice to Bob. Using our naive software on the internet, Alice would broadcast her electronic cash transaction to an online network of verifiers, who would collectively verify her transaction and transfer ownership of electronic cash.

Because the internet is anonymous, two problems arise immediately with our naive software system: Alice must prove she owns the electronic cash she seeks to transfer, and anybody can create numerous false identities to verify bogus transactions.

In other words, malicious users could create bogus transactions and fake accounts to verify their own bogus transactions. This is a type of *Sybil attack*, where a single user or colluding fraudsters create multiple identities to commit fraud or take control of the network. As you can see, this simple software is far from foolproof.

Bitcoin addressed these problems using the science of cryptography and proof of work (PoW) methods.

First, Alice can prove she owns bitcoin via cryptographic signatures by digitally signing her transactions. She doesn't need to prove her name is Alice but must verify ownership of some bitcoin via a cryptographic private key. The real-world analogy is that Alice is signing the back of a paper check and passing it to a recipient. A series of digital signatures maintains a history of ownership on the Bitcoin blockchain. We won't go into further details here, but *Mastering Bitcoin* by Antonopoulos[1] does a great job of explaining how Bitcoin works.

And how do you prevent malicious users from casting multiple votes to verify bogus transactions? As discussed earlier, proof of work was created to prevent email spam, where a user must expend computer power to send emails, thus making spam very expensive. The original Bitcoin white paper explains, "Proof-of-work is essentially one-CPU-one-vote." Bitcoin requires verifiers (called miners in Bitcoin) to solve computationally complex puzzles to verify transactions. The consensus process with proof

[1] Antonopoulos, Andreas M. "Mastering Bitcoin: Unlocking Digital Cryptocurrencies." O'Reilly Media, 2014.

of work makes it arduous for an attacker to fabricate multiple identities. The computational cost and competition among miners act as a deterrent to Sybil attacks. Miners are competing for Bitcoin rewards.

Blockchains are susceptible to the analogous political problem of the Tyranny of the Majority, in which 51% can oppress the 49%. Blockchain consensus algorithms are like pure majority rule without minority protections. A malicious individual or group might control most of the computers on the network that verify transactions for cryptocurrency rewards. Hence, they become a central authority that can verify bogus transactions or undo good transactions. By design, it takes plenty of CPU resources to manipulate the system. Proof of work isn't perfect, but a 51% attack should be extremely costly and uneconomical. The cost of bad behavior should significantly exceed the cost of good behavior.

Bitcoin was designed to transfer value: an amount of bitcoin. As mentioned earlier, Bitcoin's novel achievement circumvented the need for a third-party intermediary. Bitcoin features include transactions that require multiple digital signatures. Still, blockchain users didn't take long to want to do more complex tasks, such as creating and trading new types of cryptocurrencies or locking cryptocurrency in an escrow account, with its release subject to predetermined conditions.

Ethereum, a blockchain with a full-blown programming language, was designed to create and run decentralized applications. Programmers can write *smart contracts*, which are computer programs sent to the Ethereum blockchain for subsequent self-execution. A programmer would just set it and forget it. The blockchain becomes a self-operating machine or global computer. Smart contracts run autonomously or, like vending machines, respond to user input.

Smart contracts aren't legal contracts. Instead, they're a set of transparent rules of interaction stored on a blockchain. A smart contract sets an agreement in computer code. Any party can read the code to determine the terms and signify their acceptance by sending cryptocurrency to the blockchain's *smart contract* address. Upon receipt of an amount of cryptocurrency, the contract executes a predetermined action.

Let's explore how people conduct transactions and what makes blockchains different from what came before.

Returning to Alice and Bob, suppose she wants to buy an oil painting from him with bitcoin. This trade has two legs: Alice sends bitcoin to Bob, and Bob delivers the physical oil painting to Alice. Suppose Alice sends the payment before receiving the oil painting. The blockchain verification process assures Alice's transfer of bitcoin to Bob. However, Bob's performance occurs separately and apart from the blockchain. Upon receiving Alice's payment, Bob has an ethical and legal duty to perform his end of the agreement. However, the blockchain cannot enforce Bob's delivery of the oil painting. The delivery of the oil painting is an event that occurs in the real world and *off-chain*.

Next, consider face-to-face purchases. You typically pay cash while concurrently taking possession of goods. When you buy something online, though, you may send money and not get the goods. You take counterparty risk: the chance the other party won't deliver.

For simplicity, suppose there are two types of transactions in this world: those with counterparty risk and those without.

As an example of transactions with counterparty risk, imagine on Wall Street that a trader buying a private security might send cash via a bank wire system and accept delivery of the private security via a different banking system. It's possible for one side to deliver while the other fails to deliver. If you buy something online with bitcoin, the bitcoin part of the transaction is verified in a decentralized manner, but the other side may fail to deliver. Such a transaction has counterparty risk.

As an example of transactions with no counterparty risk, imagine on Wall Street that two parties want to trade a security but don't trust each other. The seller asks, "send me the money first," while the buyer asks, "send me the security first." Nobody wants to be first. They can use a mutually trusted intermediary, like the Bank of New York, to facilitate the trade. Both parties send their part of the trade to the Bank of New York, which will process the transaction if everything is in order. The two parties can claim they are each taking no counterparty risk, even if the Bank of New York is not perfectly riskless. However, it's as good as it gets.

The Bitcoin blockchain provides decentralized, autonomous verification of the one-way bitcoin transaction. As a result, Bitcoin-style transactions remove intermediaries but cannot remove counterparty risk for off-blockchain purchases. Regardless, that's alright if we manage the risks.

There are endless use cases for one-way bitcoin transactions. For instance, you buy something in person (sending bitcoin immediately as they standby), buy from a trusted merchant online — but you must handle disputes directly with the merchant — or send cash to a friend.

What if we wanted to execute real-world transactions on a blockchain with no counterparty risk? A blockchain obviously cannot guarantee the delivery of a real-world physical asset. However, we can write a smart contract that automates the trusted intermediary role and swaps two digital assets with no counterparty risk.

ALICE TOKENS

Alice writes a smart contract to create her own cryptocurrency called ALICE tokens. The smart contract swaps 1000 units of ALICE for one unit of ETH, the native currency on the Ethereum blockchain. Alice sends the smart contract code to the blockchain, where it is publicly viewable.

If Bob sends 1 ETH to the smart contract, he will, without a doubt, get 1000 ALICE tokens. No intermediary is needed. The smart contract verifies both sides of the trade. In the blockchain world, this is an atomic swap, which is an indivisible two-way exchange where either both deliveries succeed or none at all. Atomic swaps are building blocks for more sophisticated possibilities. We can build entire marketplaces this way.

Cute kitten videos were trendy during the early days of YouTube. A market to trade cartoon cats - crypto-kitties - became trendy early on the Ethereum blockchain. YouTube has grown far beyond cats. Whimsical cat tokens use the same blockchain technology as tokens that may represent digital certificates of ownership and authenticity for digital or real-life assets. We'll look at the uses of cryptographic tokens in the cases that follow.

From a business point of view, blockchain has two major characteristics: decentralized autonomous verification (i.e., one-way transactions) and decentralized autonomous smart contracts (i.e., two-way transactions).

- Decentralized autonomous verification.

  These are one-way transactions. One party takes an action — such as sending bitcoin to an address — while the second party may remain passive. In these one-way transactions, the crowd forms a consensus to verify transactions instead of a third party monopolizing this role. Nothing else happens on the blockchain after the transaction is verified. This is like sending cash in the mail. Once it's gone, it's gone forever.

- Decentralized autonomous smart contracts.

  In legal terms, a contract is an agreement between two parties that creates an obligation to perform or refrain from performing a particular duty. Smart contracts are not legal contracts, but the idea of mutual obligation holds true. Thus, the blockchain enforces mutual obligations. Smart contracts form the building blocks for marketplaces in the broadest sense of the word. They're a place where transactions occur.

In a face-to-face marketplace, every participant follows a set of rules. These rules determine how participants verify the authenticity of goods or cash. Auctions have rules specifying a time window for participants to inspect and verify goods, followed by bidding on the goods. Participants can shop around rapidly and return to buy the goods they want.

There are many ways to design and conduct markets. Market design is a discipline in itself. Some marketplaces have trusted intermediaries, and some do not. Companies like eBay, Uber, and Stubhub serve as intermediaries to a two-sided market, such as markets where buyers meet sellers, riders meet drivers, or ticket buyers meet ticket sellers. These companies are trusted third-party intermediaries that handle coordination and verification activities. Such marketplaces are potential applications for blockchain technology, designed to perform coordination and verification activities autonomously.

A farmers' market where participants meet face-to-face does not require intermediaries. Shoppers ask for a bunch of carrots or some other vegetable or fruit. The shopkeeper then fulfills their request in a brown paper bag as they pay with cash. This marketplace cannot be replicated on the blockchain. However, handling cash has costs and risks. Cash might be lost or stolen. Employees might skim money from the cash register. The world is already moving towards a cashless society with traditional credit cards or self-checkouts.

Here are two ways to use blockchain solutions to enhance the customer's shopping experience in our farmers' market.

First, shopkeepers can accept bitcoin as a form of payment. They hang a sign displaying the QR code of their bitcoin wallet address. Shoppers then point their mobile phone camera at the QR code to send bitcoin to a shopkeeper's bitcoin wallet address as they swap bitcoin for carrots concurrently.

Second, shopkeepers can borrow a page from how movie theaters separate ticket selling from ticket taking. A shopkeeper can issue and sell "carrot" tokens on a blockchain. Then, holders of carrot tokens would go to the shop to claim their carrots, using their tokens as a claim check. The carrot shopkeeper has safe control over money and inventory, lowers the risk of losing cash, and saves on credit card fees.

Should the shopkeeper bother with a blockchain? Would it help him increase sales or lower operational costs? It depends. He'd need to step back to see the big picture.

Part Two

Blockchain technology provides a promising new set of tools to create brand-new business models that weren't previously possible. They automate the intermediary role, threatening companies that perform coordination and intermediation.

Most business managers are paid to make incremental improvements to the company's products for existing or new customers. However, no legacy business model has ever been designed with blockchain technology in mind. As a result, attempts to append a blockchain solution to an existing business model without strategic thinking will likely fail. That's because blockchain technology is not intended for incremental improvements, as an afterthought or quick fix. Instead, blockchain is a foundational infrastructure for designing business models from the beginning.

Blockchain technology relies on advances in cryptography and networking to shift trust from a potential monopolist to a network of verifiers. To take advantage of this new technology, we must consider market trends and underlying cultural shifts.

Successful innovation occurs when technological progress meets market trends. Building blockchain prototypes is becoming more accessible as blockchain software development tools continually improve.

Figure 3

IDEAS BLOOM WHERE TECH MEETS THE MARKET WITH CULTURAL SHIFTS LEADING

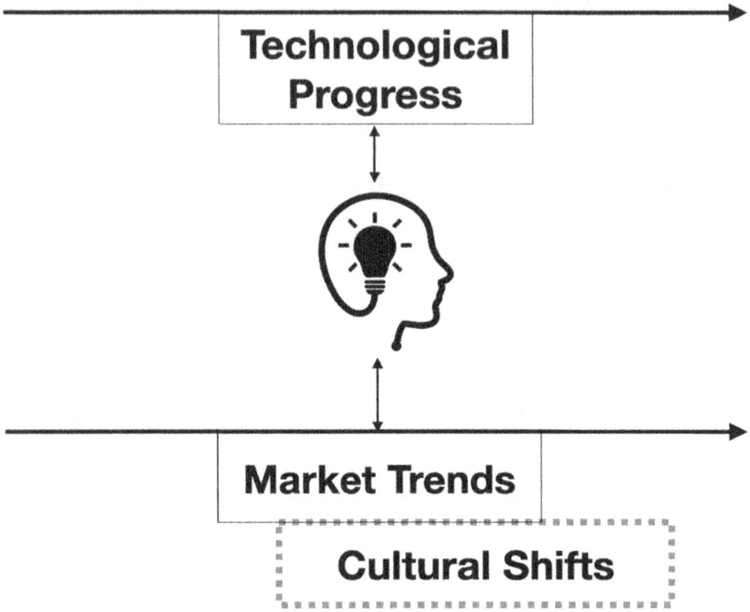

Source: Partially based on Strategic Information Technology: Best Practices to Drive Digital Transformation, 2nd Edition. Arthur M. Langer and Lyle Yorks

We can build plenty of blockchain solutions before facing actual customers. However, achieving product-market fit remains a critical challenge. The true measure of a successful product is not how much applause it receives at a blockchain conference, but rather its profitability or market share in a real-world marketplace. Technological innovations may sometimes be ahead of their time, such as when Leonardo da Vinci designed flying machines and related apparatus in the 15th century. It wasn't until five hundred years later that his drawings came to life as an airplane and a helicopter. Currently, blockchain technology is on the verge of being applied to various practical uses. Understanding how to apply blockchains will be critical in devising real-world solutions.

If we focus solely on technology while neglecting the market and cultural context of its development, we only see half the picture. Markets are observable and measurable in dollars and cents, while enthusiasm for new ideas is much harder to quantity. A hypothesis about a technologically innovative product must be tested in the marketplace. However, how do we design such a product in the first place?

Observing market trends is a good place to start. Behind these trends lie often cultural shifts that influence consumer behavior and preferences. Cultural shifts are qualitative and often difficult to measure or forecast. Part of the creativity in designing new products entails melding technology, markets, and culture.

For example, our society appears to be moving towards increasing skepticism and declining trust in large corporate, governmental, or religious institutions. Such a world may be ripe for decentralization, where two parties can transfer money without going through a large institution. Blockchain solutions may include voting systems, charitable donation verification, supply chain tracking, identity verification, and many others.

The expectations around new technologies are often inflated in the short term, only to deflate in the long term. During the internet bubble craze of the late 1990s, the mere mention that a publicly traded company had developed a website could trigger a surge in its stock price.

Some of the same zeal drives the latest waves of technological innovation, such as artificial intelligence and blockchain technology. The booming internet companies of the late 1990s to early 2000s integrated the internet into their business model rather than simply append a website to an existing one.

As new technological innovations emerge, firms must adapt their business models and, more specifically, their revenue models. Early internet firms discovered ways to monetize personalized and targeted advertising while profiting from personal data—capabilities that did not exist before the internet. As blockchain technology destroys the market power of intermediaries, firms must discover new pathways for gaining market power.

Figure 4

CREATIVE DESTRUCTION, CREATIVE CONSTRUCTION

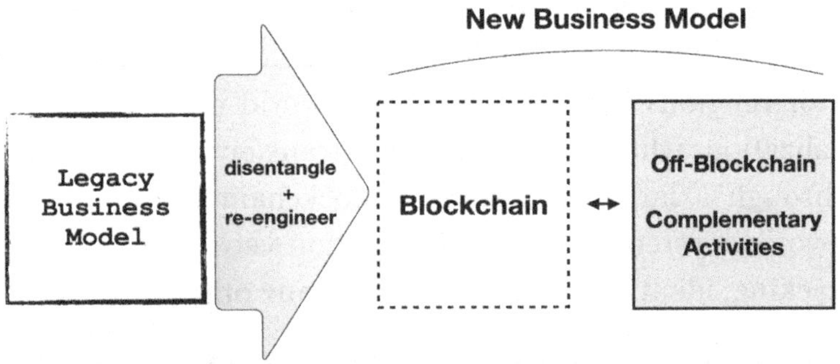

Blockchain technology is available to everyone. Ultimately, blockchains will be everywhere as a foundational infrastructure for any business. Decentralized blockchain applications will operate as layers on top of blockchains, with most company processes or complementary activities taking place off the blockchain.

Blockchain Layer	Description
Layer 1 — Foundational	Blockchains, such as Bitcoin and Ethereum
Layer 2 — Scaling	Extensions to Foundational Layer designed to support growth and increase transaction speed (e.g., Lighting Network for Bitcoin).
Application Layer	Decentralized Apps that use smart contracts and require business domain knowledge (see Part III, Case Studies).

We can break down all things blockchain into three layers, from the bottom up: Foundational (i.e., Layer 1), Scaling (i.e., Layer 2), and Application.

> *Layer 1* is the first layer providing foundational infrastructure. It provides the essential services of a blockchain network. Examples include Bitcoin, Ethereum with smart contracts, as well as hundreds of other blockchains.
>
> *Layer 2* is the second layer in the infrastructure designed to support the growth or scale of blockchain. For example, Bitcoin operates at around ten transactions per second, while the traditional credit card Visa network runs at six-thousand transactions per second[1]. This layer overcomes these limitations, preparing blockchain technology for mass adoption. A compelling scaling solution is the Lighting

[1] "What Is Lightning Network?" Bitcoin.com, https://www.bitcoin.com/get-started/what-is-lightning-network/.

Network, a secondary network that allows near-instant, high-volume bitcoin micropayments.[2]

The Application Layer or decentralized app — which is the focus of the case studies in this book — is where business domain knowledge comes into play. This is where the digital and real worlds meet. This includes off-blockchain or complementary activities. Examples include music distribution, land titles and related regulations, supply chains, and the last-mile problem.

To use blockchain technology's capabilities effectively, a business manager must start with strategic thinking. Strategy is not just about planning. As boxing champion Mike Tyson once said, "Everyone has a plan until they get punched in the mouth." Strategies are those plans that specifically focus on the actions and responses of competitors.

Most businesses perform a bundle of activities. Managers must introspectively assess the strengths and weaknesses of their firms in a post-blockchain environment relative to competitors. Business managers need to assess where blockchain could and should be used, considering that competitors can also use it and that not all activities match the technology well. To start our assessment with the end in mind, we must think about how to gain or destroy market power in a world where blockchains become omnipresent.

[2] Poon, Joseph, and Thaddeus Dryja. The Bitcoin Lightning Network: Scalable Off-Chain Instant Payments. 14 Jan 2016.

Chapter Three

Market Power

Every participant in a political or business relationship has relative power, whether explicit or implicit. Political power exists on a continuum between total anarchy and absolute dictatorship. Market power also exists on a spectrum: from the romantic ideal of perfect competition to a monopoly with no competition.

Figure 5

Market Power Exists on a Spectrum

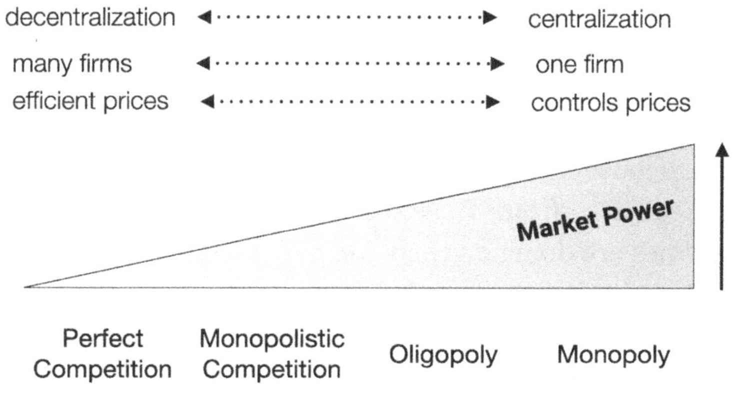

As explained earlier, blockchain technology can undermine both political and market power by automating the intermediary role. Blockchain technology is part of the new economic structure in what German economist Werner Sombart called creative destruction[1] — the process of industrial mutation that continuously revolutionizes the economic structure from within, incessantly destroying the old, and creating a new one.[2]

We've seen claims about the democratizing influence of technology. The internet gave billions of people worldwide instant access to information. Anyone can launch their own website or write a blog. Many businesses, from mom-and-pop shops to corporations such as Barnes and Noble bookstore, launched their websites as a front-store display window. However, information overload makes navigating and filtering for reliable content difficult for individuals.

The internet has become an essential infrastructure that is relied upon for day-to-day activities. Forbes reports there are now over a billion websites[3] — about one site for every seven people in the world — but we can only remember or use a small handful. Firms with dominant market power have come to rule the internet.

The internet has given rise to centralized corporate intermediaries. These corporations include social media platforms and payment companies. We are obliged to use the services of these platforms to buy books online (i.e., from Amazon), purchase sports tickets (i.e., from Ticketmaster), or see what our high school buddies are doing on their vacations (i.e., on Instagram). As a result, such firms gain privileged information and market power. These companies don't compete with each other, but

[1] Reinert, Erik. "Creative Destruction in Economics: Nietzsche, Sombart, Schumpeter (with Hugo Reinert)." The Visionary Realism of German Economics: From the Thirty Years' War to the Cold War, edited by Reinert Kattel, Anthem Press, 2019, pp. 385-412.

[2] Creative destruction is often attributed to Joseph Schumpeter. Still, while he certainly popularized the term, it was initially coined by Werner Sombart, who was influenced by the works of Friedrich Nietzsche. Going back in time, the process of creation and destruction plays a central role in Hinduism, the religion which inspired Nietzsche's Erzieher (educator) Arthur Schopenhauer.

[3] Haan, Katherine. "Top Website Statistics for 2024." Forbes, 4 June 2024.

rather dominate distinct markets. Competition is eroding among the top firms.

We're interested in this notion of market power and competition because thriving competition is a basic tenet of a well-functioning free market economy. And competition, we're taught, is good for consumers and society. Competition lowers consumer prices, expands consumer choice, and encourages companies to innovate and invest. Competing firms soon notice if anyone sells a product well above its production costs, earning juicy profit margins. They'll want to enter a very attractive market. To be competitive, new entrants must charge a little less for the same goods. Competition leads to a reduction in the markup of goods.

Figure 6

RISING MARKUPS SUGGEST DECLINING COMPETITION

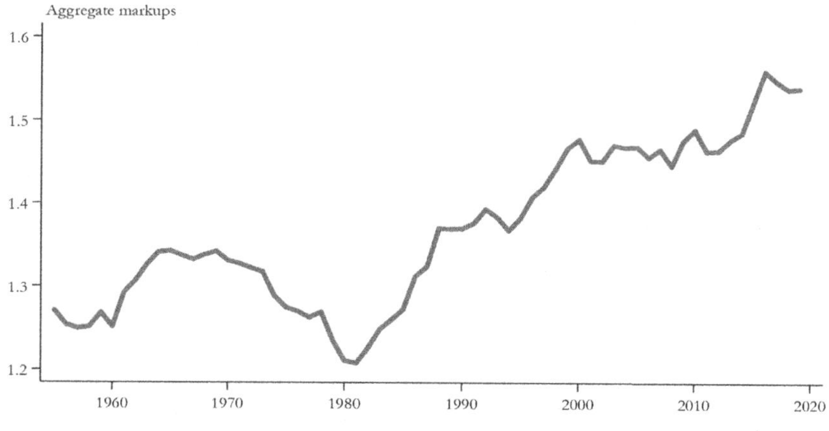

Source: The Rise of Market Power and the Macroeconomic Implication, Jan De Loecker, Jan Eeckhout, Gabriel, Unger.

What if we observe consistently high markups and no new entrants seeking to compete? Why don't other companies jump in to compete? High markups suggest the absence of competition, indicating that some companies have market power — the ability to set prices above the level that would prevail under competitive circumstances.

According to studies by economists Jan De Loecker of Princeton University and Jan Eeckhout of University College London[4], average markups — defined as the ratio of price to marginal costs — have surged since 1980 (See Figure 6). De Loecker and Eeckhout analyzed the balance sheets of publicly listed U.S. companies from 1950 to 2020. The average markup was 18% in 1980, but by 2020 it was 67%. The study uses markups as a proxy for market power. Such excessive markups should attract new entrant competitors that would diminish markups. However, the absence of new entrants is evidence of rising market power.

Consider that in 1980, the biggest companies in the U.S. had formidable competitors, but not so in 2020. In 1980, the biggest companies were in the oil and automotive industries. The oil industry had heavyweights Exxon, Shell, and Mobil cohort companies. The auto industry had General Motors, Ford, and Chrysler. Large firms competed with one another. Consumers had choices and could shop around among many gasoline stations and car makers.

A large company without a rival at the time was American Telegraph & Telephone (AT&T). It was eventually deemed a monopoly and broken up because of antitrust laws. Note that technological innovations are often several steps ahead of the legal system.

[4] Loecker, Jan De, and Jan Eeckhout. "The Rise of Market Power and the Macroeconomic Implications." National Bureau of Economic Research, 21 Aug 2017.

In 2020, the biggest companies were in separate lanes: Facebook, Amazon, Apple, Netflix, and Google. Who are their rivals? What easy choices do consumers have?

What prevents a firm from raising prices, selling our private data, and gaining market power without competition? Firms naturally strive to become monopolists, like a football team striving to win championships every single year. We'd all do the same if we could. However, the irony is that competition may lead to no competition if a few firms gain overwhelming market power. How have companies been able to maintain market power in a competitive marketplace?

The Rise of Intangible Assets

Economists Nicolas Crouzet and Janice C. Eberly suggest that rising market power is associated with intangible assets.[5] Intangible assets are non-physical assets, such as software, patents, and trademarks. Companies with more intangible assets on their balance sheets have tended to show higher gains in market power in the 21st century.

The two largest firms in 1980, Exxon and General Motors, each operated with more than 300,000 employees and invested heavily in physical plant and equipment. Conversely, by 2020, the two largest firms, Apple and Microsoft, had under 150,000 employees each and invested heavily in intangible assets.

Consider a firm expanding to a new location. An industrial firm must spend funds on a physical plant and equipment for a new assembly line (e.g., robotics). The moving parts of those robotic arms wear out and occasionally need to be replaced. However, the same software program running the robotic arms can run

[5] Crouzet, Nicolas and Eberly, Janice, Understanding Weak Capital Investment: The Role of Market Concentration and Intangibles, NBER Working Paper No. w25869, May 2019.

countless times. Software incurs an initial fixed cost, but it can run and be cloned virtually endlessly. The same software code can run in China and the U.S. simultaneously.

Now, let's try to gain market power with tangible assets such as physical plant and equipment. If a particular market has sufficiently attractive opportunities, entrepreneurs or companies should be able to find investors to raise enough capital to buy the tangible assets needed to enter the attractive market and compete, eventually lowering markups. Makers of tangible assets generally will sell to anyone. Hence, new entrants can usually buy all the tangible assets they need.

Figure 7

Can't Buy Me Intangibles

Tangible vs. Intangible Assets for S&P 500 Companies, 1975-2018

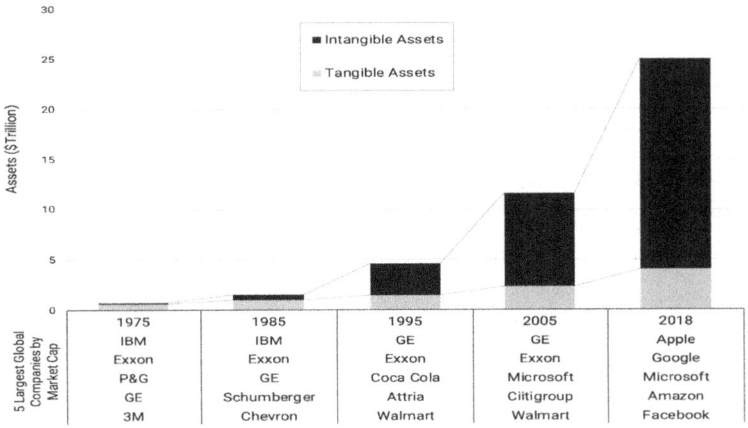

Source: UCLA Anderson Review[6]

[6] Brown, Monika. "Boom of Intangible Assets Felt across Industries and Economy." UCLA Anderson Review, 19 Apr 2023.

However, new entrants cannot easily acquire intangible assets. For example, to compete with Facebook, you'd need to acquire their customers—an intangible asset. To compete with Google, you'd need plenty of private data. With Pfizer, you'd need pharmaceutical patents. You can't easily buy intangible assets.

Hence, tangible assets today rarely are the source of market power. There are some exceptions, as in the financial services industry, where large banks with piles of cash, like J.P. Morgan, inspire confidence to withstand economic shocks.

In summary, an intangible asset, such as software, a patent, or a trademark, can be used simultaneously by multiple units at a firm while being protected from imitation by the force of law. Tangible assets are more accessible to buy than intangible assets. Unlike tangible assets, intangible assets are evidently more likely to produce returns above their marginal cost. This advantage can lead the firm to win a high market share and move closer to beating and eradicating the competition.

Firms gain market power by putting their tangible and intangible assets to work for them. A firm with market power must either do something that rivals cannot do, or do what rivals can do, but do it better.

Chapter Four

Barriers to Entry

Bruce Greenwald and Judd Kahn (*Competition Demystified: A Radically Simplified Approach to Business Strategy, 2007*) argue that on a level playing field, many competitors have essentially equal access to customers, suppliers, new technologies, and other cost advantages.[1] In a market open to competitors on equal terms, competition erodes the returns of all to a uniform minimum.

Rivals will copy anything that another does to improve its position. For a firm to earn above-average returns, it must do something competitors cannot. Therefore, to achieve superior returns or superior market share, which we call superior performance, a firm must either operate more effectively than competitors or establish and maintain a *barrier to entry*.

Barriers to entry are legal, technological or market forces that discourage or prevent new entrants from joining a market. Barriers to entry can range from the simple and easily surmountable (e.g., the cost of a license) to the extremely restrictive (e.g., a patent or an onerous government contracting process). In some cases, barriers to entry may limit competition to a few firms or lead to a monopoly.

Entrepreneurs Eric Reis (*The Lean Startup, 2011*) and Ash Maurya (*Running Lean, 2022*) have used the phrase "unfair

[1] Greenwald, Bruce C., and Judd Kahn. Competition Demystified: A Radically Simplified Approach to Business Strategy. Penguin, 2007.

advantage" as something that cannot easily be copied or bought. Similarly, "economic moat" and "secret sauce" describe the same concept. They help explain the basis for a firm's market power.

A business could be the first to use blockchain technology, gaining a *first-mover advantage*. However, anyone could also use blockchain technology. First-mover advantage is soon eroded if others can easily copy and catch up.

Figure 8

DISENTANGLING TO FIND CAUSES OF SUPERIOR PERFORMANCE

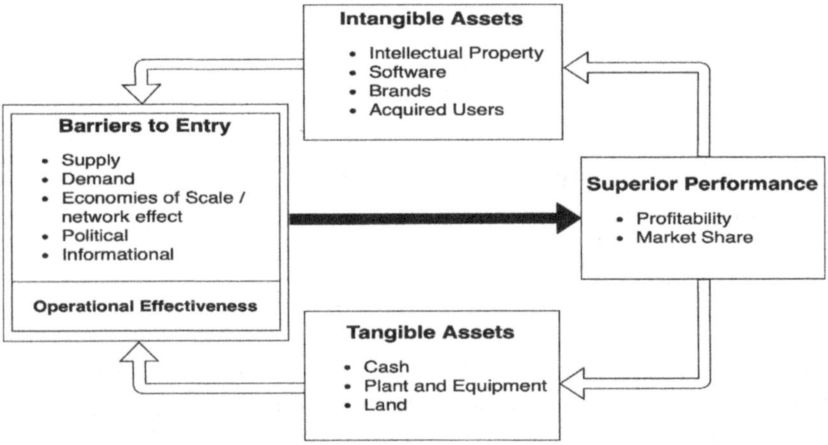

Source: Partially based on Greenwald and Kahn (2007) and McCarthy, Rouse, and Serban (2015)[2]

An effective business model with blockchain technology should exploit your unique advantages relative to competitors. What is your special access to customers, supplies, or information?

[2] McCarthy, N., B. Rouse, W., & Serban, N. (2015). Disentangling Competitive Advantage and Superior Performance and Their Roles in Enterprise Transformation. Journal of Enterprise Transformation, 5(2), 113–140.

Employ blockchain technology and bar others from enjoying your superiority. Start with that goal in mind.

Greenwald and Kahn (2007) define the most important barriers as supply, demand, and economies of scale. Let's review and break down each barrier in detail.

Supply Advantage

Supply advantage is based on having lower costs, special access to supply, critical inputs (e.g., raw materials), or intellectual property protections (e.g., patents or trademarks). For instance, you can get raw materials cheaper than competitors. Intellectual property, such as patents or copyrights, is a government-granted monopoly that excludes competitors from the market by force of law — even if only for a limited number of years. Intellectual property is becoming an increasing part of a firm's competitive advantage.

For example, Bitcoin miners in Iceland have a cost advantage because they can access cheap electricity to run their computers, and the cold climate lets them save money on air conditioning/cooling systems. Computers that run continuously risk overheating — but less so in Iceland.

Another example is Amazon. It has a major cost advantage over brick-and-mortar stores because it operates without the high cost of retail outlets. Critically, much of its inventory is either on-demand or on consignment. As a result, Amazon has less money invested in inventory despite being primarily a reseller. Amazon also gets paid upfront. It generates cashflow sooner than stores that must perpetually fund and restock their local inventory and then wait to sell and get paid while covering rent and other expenses.

One more example would be pharmaceutical companies. They enjoy a supply advantage in the domain of intellectual property. Their patents are temporary monopolies that allow for huge markups. A drug might cost pennies to manufacture but sell hundreds of times more because their customers need the product and no one else can produce it.

Demand Advantage

Demand advantage is about having captive or loyal customers — or high switching costs. Companies with a demand advantage could be experts at establishing long-term relationships with their customers or making it very difficult for them to leave.

An example would be Oracle Corporation. Its database software requires significant upfront commitments from users in terms of time and money. Changing to another platform would make clients feel they wasted time and money customizing their Oracle system. Plus, it can take plenty of more time and effort, such as retraining their staff, to migrate to something else.

Salesforce, Inc. provides customer relationship solutions, customizing and embedding its software into its clients' workflow and operations, requiring extensive staff training. The company convinces clients to make a significant investment in time and money, making it unlikely that clients will switch to a competitor later. Further, if a competitor offers some new feature, Salesforce needs only to do the same, making it easier for clients to upgrade than to switch systems.

Economies of Scale Advantage

This is the advantage a firm enjoys due to its scale of operations. To understand this better, let's divide it into two parts: supply-side and demand-side.

Supply-side economies of scale advantage is the traditional meaning of economies of scale - a situation where a large scale of operations results in lower per-unit cost. For example, Walmart is a single large buyer with bargaining power over many smaller suppliers. As a result, it receives volume discounts to pass on to its customers. Then, Walmart attracts more customers, thereby becoming larger and increasing its advantage in the marketplace.

Demand-side economies of scale advantage is also called the *Network Effect*. It's increasingly significant in the interconnectivity of the internet age. Network effects are based on the notion that the value of a service increases as more customers join their network. The classic example was the phone company. A more recent example is Meta's Facebook. Users find more value as more of their friends join the network. A startup could build another system similar to Facebook, offering a better interface or other enticing features. However, they would have a tough time luring away Facebook's customers, whose friends are all on Facebook.

The proliferation of mobile phones, internet access, and the advent of blockchains inspire us to single out and expand upon two more barriers to entry alluded to by Greenwald and Kahn (2007): political and informational ones. Such barriers to entry have grown in significance since 2007.

Political Advantage

Political advantage is conferred via explicit government protection, such as subsidies or special access to policymakers. It could also result from a quid-pro-quo following lobbying expenditures or political donations.

Some firms may never talk to a government official but are better at navigating bureaucracy. They have a knack for anticipating new rules or regulations and developing products that exploit them ahead of competitors.

Other firms may be best at lobbying for rules, regulations, or standards that are favorable to themselves and unfavorable or difficult to others.

Yet other firms push the limits of legal or regulatory compliance, risking occasional fines or penalties merely as the cost of doing business. Startup entrepreneurs, driven by the need for growth, have more incentives to take legal or regulatory risks than established firms with existing business interests at stake.

Some examples of these types of firms or situations might be useful.

Amazon negotiates with local governments for favorable customized tax breaks unavailable to competitor businesses. This two-tier tax system creates a situation where Amazon pays lower taxes and competing businesses pay higher taxes. In a sense, local government officials intervene to pick winners and losers, supposedly in the interest of their constituents, to bring more jobs to their community. However, having several mid-size companies offering similar benefits to the community might diversify and strengthen the local economy. A fair tax system lets the free market decide.

Another example is traffic enforcement camera makers. Many coordinated with local governments to reduce yellow light times for profit[3]. Drivers are accustomed to about 4 seconds. Shortening the time increases the number of tickets for traffic violations, thereby increasing revenues for the municipality and profit for the company — all at the expense of drivers.

[3] Zurschmeide, Jeff. "Why Red-Light Cameras Are More about Money than Safety." Digital Trends, 13 Mar 2016.

The Federal Reserve Bank provides cheap financing to big banks and bailouts when markets crash. Many economists support this system as a way to manage monetary policy for price stability. However, it arguably gives some banks an unfair advantage over non-bank lenders.

National Football League owners have been known to lobby for the construction of taxpayer-subsidized football stadiums. However, a Federal Reserve research report references a 2017 poll in which 83 percent of the economists surveyed agreed that "Providing state and local subsidies to build stadiums for professional sports teams is likely to cost the relevant taxpayers more than any local economic benefits that are generated."[4]

Informational Advantage

Informational advantage is based on having secret or privileged information, such as insider information, private personal data, and transaction or transaction flow data. This means having information that others do not (e.g., private user data), which companies can sell or monetize for profit.

Artificial intelligence algorithms can process data to predict or influence your behavior. There is little academic research to build upon for informational barriers. Still, authors such as Shoshana Zuboff (who popularized the term *surveillance capitalism*), Yuval Noah Harari, and Jaron Lanier have raised the alarm around the growth of data in our lives and how companies monetize private data.

Let's consider some examples:

In the U.S., internet service providers may lawfully profit by selling your browsing data.

[4] Wolla, Scott A. "The Economics of Subsidizing Sports Stadiums." Economic Research, Federal Reserve Bank of St. Louis, May 2017.

Credit card issuers, banks, and software vendors acquire personal identifiable information (PII) to increase their credit offers to customers and sell personal data to data brokers, a multi-billion dollar industry.

Amazon is lobbying against restrictions on private data. They may be motivated in this area because of their Alexa device, which listens to what you're saying in your home.

Company policies may claim to sell only so-called anonymized data by omitting your name from the data they sell about you. A study at Carnegie Mellon University found that 87% of the population of the United States can be uniquely identified by only knowing their gender, 5-digit zip code, and date of birth.[5] Privacy is no longer very private. Data brokers sell data to big investors, who might use so-called anonymized data to cross-reference credit reporting and mortgage data to predict mortgage defaults based on late credit card payments.

OPERATIONAL EFFECTIVENESS AND FIRST MOVERS

If a firm achieves superior performance but we can't determine at least one barrier to entry, we could surmise that such a firm must have achieved operational effectiveness. We should bear in mind that we mean effectiveness, not efficiency. The distinction between these two notions is that effectiveness is all about doing the right thing before doing things right.

We should find the true barrier to entry. It might be a closely held secret. Firms are incentivized to hide their barriers to

[5] Sweeney, Latanya. "Simple Demographics Often Identify People Uniquely." Carnegie Mellon University, Data Privacy Working Paper, no. 3, Pittsburgh, 2000.

entry, especially political or information advantages. They may have more than one barrier to entry. For example, a network effect might be erratic, like a fashionable restaurant – until the next trendy restaurant opens. A network effect plus sticky customers is a much stronger advantage.

First-mover advantage is a jump start to gain an early lead against the competition. However, an early lead may not be sustainable. A firm must capture the benefits of being first or risk losing to more formidable competitors. The first mover must eventually erect a sustainable barrier to entry. Let's consider some examples of this strategy:

Establish your product as an industry standard. This creates sticky customers for your product, solidifying a Demand Advantage. Motorola embraced its longstanding role as an electronics industry trailblazer in global standards. Do an ostensibly good deed for the world by pushing for your own standards.

Rapidly grow a Network Effect and combine it with captive customers. A typical strategy is rushing to capture market share before new entrants join. Then, what keeps new entrants from stealing your customers? Your customers need to be held captive. Solidify the holy grail of Network Effect plus Demand Advantage.

Create a new exotic financial product ahead of the competition. Charge high fees when you're the only one offering a new product. Wall Street banks earn higher fees from new exotic financial products until the novelty wears out. Then fees compress to normal. This is product differentiation in marketing.

For example, Myspace had a first-mover advantage over Facebook, and AltaVista had a first-mover advantage over Google. Being first may or may not lead to a long-term barrier to entry.

Instead, the first mover may only prove an idea's viability, trailblazing a clear path for new entrants to take over.

Barriers to entry rightfully conjure a huge, immovable boulder. In practice, it takes time to erect such barriers when a business is new and growing. Speed is essential, but it can work against you. Trailblazers must be cognizant of copycats. Suppose a military commander attacks and captures a fortress by exploiting a weakness in a rear wall. The attacker becomes the new occupier who must defend the same vulnerable rear wall against future attackers. The perched occupiers must erect a barrier to entry to stay in power.

Facebook enjoys the combination of Network Effect and Demand Advantage (e.g., sticky customers). Uber enjoys a Network Effect but not a strong Demand Advantage - its customers may not be so sticky. An Uber copycat startup has a chance of stealing Uber customers because riders care little if their friends also ride Uber. Facebook users are sticky or captive, but Uber users can easily switch to Lyft or a potential decentralized solution.

There are many strategies for success. What matters is that each company examines its strengths and weaknesses, and assesses how best to exploit them.

Part Three

This part contains six practical case studies, each from the point of view of a specific character, someone who uses a blockchain solution to solve a real-world problem.

Each case starts with a problem and current situation in the music industry, a supply chain, Uber-like mobile taxis, real estate finance, online privacy, or gold-backed ETFs. Our underlying themes are that blockchain technology can destroy political or market power to the benefit of end-users. We posit that central authorities, such as music labels or central banks, have power, but our protagonist does not.

Given a problem, what exactly do we want? Next, each of the six cases defines a wish list in the form of very detailed Success Criteria, a checklist that tells us whether the subsequently proposed solution works.

The meat of each case is the solution. Each case provides a breakdown of the design, planning, and execution of a solution. This section is akin to high-level IT specs. It serves as a discussion document for a business and IT manager meeting. We use sequence diagrams.

Sequence diagrams illustrate the interactions and message flows among different participants or components of a system over time. Vertical lines - called lifelines - represent the participants or components involved in the scenario. The messages exchanged among them are shown as horizontal arrows or lines.

The diagrams read from top to bottom, representing the chronological order of interactions.

Given our solution, we expect risks. We address ways of mitigating them. For example, in all cases, blockchain solutions are written on top of and depend on the reliability of an underlying blockchain. Following are the six cases in this part.

SIX CASES

Digital Jukebox

For artists who receive low payouts for their music, we'll build a decentralized solution so they can transact directly with their fans.

Trading Bills of Lading

For small suppliers and buyers who get poor trade financing terms, we'll build a decentralized solution so they can trade goods in transit and advance their cash flow positions.

Car-Sharing Service

For drivers who receive only 50% of the fare as payouts from ride-sharing companies, we'll build a decentralized solution so they can transact directly with their riders and shrink the margin between fares and payouts.

Real Estate Finance

For real estate developers struggling to kick start a new project, we'll build a decentralized financing system so they can resolve the collective action problem by transparently validating demand while raising capital.

Monetizing Private Data

For users who are concerned about privacy on the internet, we'll build a decentralized tool so they can control and monetize their own private data.

Gold-Backed Tokens

For anyone who likes gold, we'll launch a decentralized gold-backed token so they can transact in gold instantly with anyone in the world.

❦

Notice how each of these cases puts power back into the hands of the original creator or innovator, and returns data and control to the individual. This is how blockchain — based on a string of alphanumeric characters that reside in code and traverse a digital network — can be an agent for change in the real economy.

Case One

Digital Jukebox

Until recently, any artist who wanted to succeed in the music business knew there was only one path to success: being discovered, signing a contract with a corporate label, and becoming a star. All of that depended on music executives' whims — and finances.

However, while a small minority of musicians achieve fame and financial success, this model allows record labels to act as gatekeepers, ensuring that most do not — or barring others from competing. The life of an artist is one of relentless hard work and very little agency, with no guarantee of success. For the few that do get a record contract, most capture only 12% of music revenue, according to Citigroup[1].

Nowadays, streaming services like Spotify have disrupted the music industry and put millions of songs at the fingertips of listeners. Still, Spotify's four largest music suppliers — Universal Music, Sony, Warner, and the independent coalition Merlin Network – combined make up 77% of total music streams, and 99% are from Spotify's own curated list.[2]

What if a musician could bypass the industry machinery? How can blockchain technology support greater autonomy and generate a higher return on the same amount of work?

[1] "Putting the Band Back Together." Citi GPS: Global Perspectives & Solutions, Citigroup, August 2018.
[2] Spotify 2021 Annual Report.

In this case, we'll explore how a musician, Aria, can use non-fungible tokens (NFTs) to sell music directly to her fans. This allows her to spend more time and energy creating music and building her brand, instead of seeking a corporate deal, thus losing her independence.

In this model, Aria collects the lion's share of royalties and retains the rights to all her music. She has the choice of whether to sell these rights.

The dream of a celestial jukebox[3] allows everyone in the world to access all the music content ever created from anywhere, at any time. Ideally, artists like Aria have a one-to-one relationship with their fans, delivering music directly to them and keeping the bulk of the revenue.

Current Situation

A traditional record deal is a complex contract that effectively functions as a loan to the artist, repaid from future music sales. Artists start collecting money only after their debt is paid back. "It's essentially like indentured labor, where nine times out of ten, the artist will remain in debt to the record label for the initial upfront advance that the record label paid them and they're never paid off. And they remain tied to that record label with a relatively small degree of control over their commercial approach, marketing approach and no ownership of the copyright." says Mark Mulligan, Media Analyst and Founder of Midia Research in the Financial Times documentary film, "How record deals work and make money in the music industry."[4]

[3] Goldstein, Jacob. "Copyright's Highway: From Gutenberg to the Celestial Jukebox." Stanford Law and Politics, 2003.
[4] Financial Times. "How Record Deals Work and Making Money in the Music Industry | FT Film." YouTube, 14 Apr 2021, www.youtube.com/watch?v=XIgcPomfnoc.

Record labels may also offer a "360-degree deal," in which an artist pays the record label a share of all revenue sources, including touring, merchandising, licensing, publishing, and endorsements.

Figure 9

RECORD LABEL GETS 80% OF REVENUE

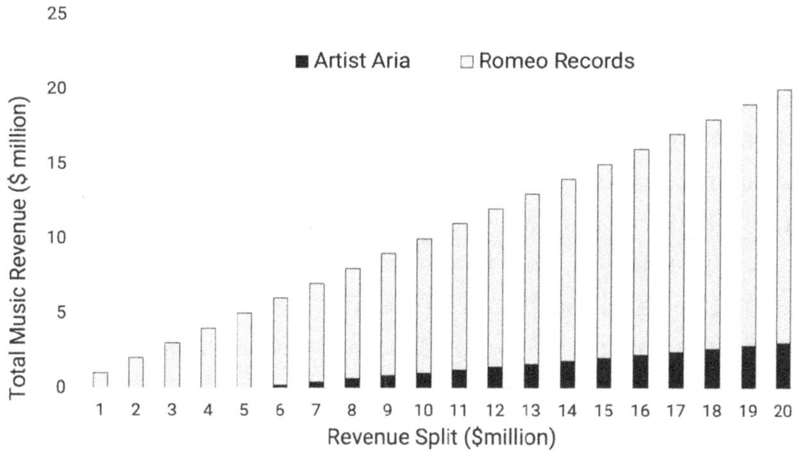

For example, suppose our Aria signs a record contract with Romeo Records label. The terms are as follows:

- Romeo Records pays Aria an advance of $1,000,000
- 80/20 Royalty Split – Romeo Records gets 80% of royalties, and Aria gets 20%
- Romeo Records recoups recording costs out of Aria's share
- Romeo Records owns the master sound recordings.

Under these terms, Aria's 20% royalty split from the first $5,000,000 in sales will pay back her $1 million debt.

Record companies are compensated for taking on the risk of signing new acts and providing the resources to produce, promote, and distribute music. Aria may otherwise remain unknown without a record deal. However, the economic dynamics began changing in the early 2000s with the shift from physical to digital music, and the emergence of social media and online music distribution platforms. Technological progress creates new ways for artists to do-it-yourself (DIY) without a record label, and deliver music directly to their fans.

Nowadays, artists have the tools to produce, promote, and distribute their music. Tech-savvy individuals can set up sophisticated recording studios at home. International DJ Calvin Harris produces his music using Apple's Logic Pro software.[5] On the self-promotion and branding front, then-unknown rapper Megan Thee Stallion began uploading freestyle videos on social media as a Prairie View A&M University student.[6] A then-unknown Billie Eilish uploaded her songs on the music-sharing website SoundCloud.

SoundCloud and YouTube offer their content creators a cut of advertising revenue. However, content creators must take the price established by powerful corporations. As with most content creators, artists have weak bargaining or market power.

Ideally, Aria should focus her energy on her passion, which is to make music. Aria has recorded ten original songs and has built a dedicated fanbase, many of whom attend her regular performances at the Rock-A-Bowl club. Despite her talents, she returns to her 9-to-5 office job, not ready to pursue music full-

[5] Watley, Alex. "What DAW Does Calvin Harris Use? Discover the Music Production Software Behind His Hits." AC3FILTER.NET, 2 Sep 2023.
[6] Howard, Nandi. "Meet the Electrifying Megan Thee Stallion." TheFADER.com, 30 Aug 2018.

time. Aria doesn't necessarily dream of signing a record contract, which provides a small chance of a big payoff and a big chance of no payoff. Instead, Aria would be happy with moderate success, making a modest living doing what she loves. Smaller, talented artists often go unnoticed by major labels, who may prefer to invest in acts they deem to have greater potential for high rewards.

Pragmatically, Aria must build a brand, be part of a community, and build a close relationship with her fans. A brand helps to create sticky customers, or what in business strategy terms is called having a "demand advantage." In other words, it is what makes them unique.

Songwriters like Aria may also sell their song rights to investors. In the real world, Hipgnosis Song Fund, a UK-based investment company, acquires song catalogs built around proven hit songs by best-selling artists such as Shakira, Ed Sheeran, and Bruno Mars. A blockchain solution, though, could start with small artists that large labels ignore.

Success Criteria

Our objective is to provide an easy, do-it-yourself way for Aria to get paid by transacting directly with her fans without needing a third-party intermediary.

Our success criteria form a checklist of requirements for our solution.

- **Establish a peer-to-peer relationship between Aria and her fans.**

  We seek to simulate the face-to-face experience of Aria performing live before an audience and immediately collecting tips or payments at the door. The solution must have an easy user interface for Aria to upload her songs and make them available to fans.

- **Aria owns her master recordings.**

  As such, song downloads or streaming payments go directly from fans into Aria's wallet.

- **Aria gets paid first.**

  Aria may route money to her own service providers, such as a manager, promoters, and influencers. This flips the business model whereby the artist is the first, not the last, to get paid.

- **Aria may sell her master recordings.**

  Aria can cash out of her entire music catalog. She may also sell her master recordings with restrictions. For example, an artist like Neil Young might want to avoid having his music played at political rallies. Aria may sell exclusive rights to individual collectors. For example, a collector may buy rights to a live concert recording. Fans may want their favorite artist to sing "Happy Birthday" to them to personalize their experience with the artist's music.

Solution

For artists who receive low payouts for their music, we'll build a decentralized solution so they can transact directly with their fans.

We disentangle the services of a record label, with artists like Aria in control. Aria builds a community, uses Non-Fungible Tokens (NFTs) to sell creative work, and then outsources marketing and advertising services.

There are several players in our solution. There's Aria, a creative artist who composes and records songs using Native Instruments software. She may also use the services of companies that work with independent artists (e.g., The Orchard, which specializes in marketing and advertising).

Next is Buyer Bob, a fan of Aria's music. There's also Investor Iris, who wants to buy Aria's master sound recording.

Let's review NFTs and define Oracles before proceeding to the solution.

NON-FUNGIBLE TOKENS

A non-fungible token (NFT) is a digital representation of ownership for a unique asset, serving as both a title of ownership and a certificate of authenticity. The non-fungible nature of an NFT means it is unique and not interchangeable with any other token.

Technically, an NFT is cryptographic computer code recorded on a blockchain to prove ownership and authenticity of a unique asset. Most other blockchain tokens are fungible or interchangeable. For example, two different bitcoins are primarily alike, and being the holder of either bitcoin would give the owner the same value or rights as the other. However, an NFT representing an event ticket for a specific seat (e.g., Row A, Seat 1) is non-fungible because it grants unique rights.

An NFT can represent almost anything, provided it is a unique asset. NFTs often contain links to off-blockchain data, such as to the location of a video, image, or music file.

ORACLES

In Greek mythology, an oracle is a special person who communicates with the Gods, passing along wisdom and advice. In the context of blockchain technology, oracles are mechanisms for blockchains to connect with external data. This data goes to smart contracts, which run autonomously on the blockchain. Ideally, oracles should be trustless and avoid a single point of failure or control. Data providers don't need to trust each other to reach a consensus.

For example, we can write a smart contract to insure against a category four hurricane making landfall in Florida in the next twelve months. If a qualifying storm occurs, real-world weather information must reach the smart contract to trigger a payoff. The benefits of decentralization are lost if the trigger relies on a single individual providing the data, as they might have a vested interest in the outcome. The decentralized way to trigger the payoff involves a group of people or entities called oracles, who each submit their vote to reach a group consensus. Now, it takes a group decision to trigger the smart contract, eliminating the need to trust a single person or entity.

The oracles in this case are groups of individuals who vote on copyright claims against the songs. Blockchains know nothing about the real world, including the weather, copyright laws, and the outcome of sporting events. For such situations, blockchains need Oracles to obtain reliable external data.

Figure 10

ARIA CREATES AN NFT AND COLLECTS ROYALTIES

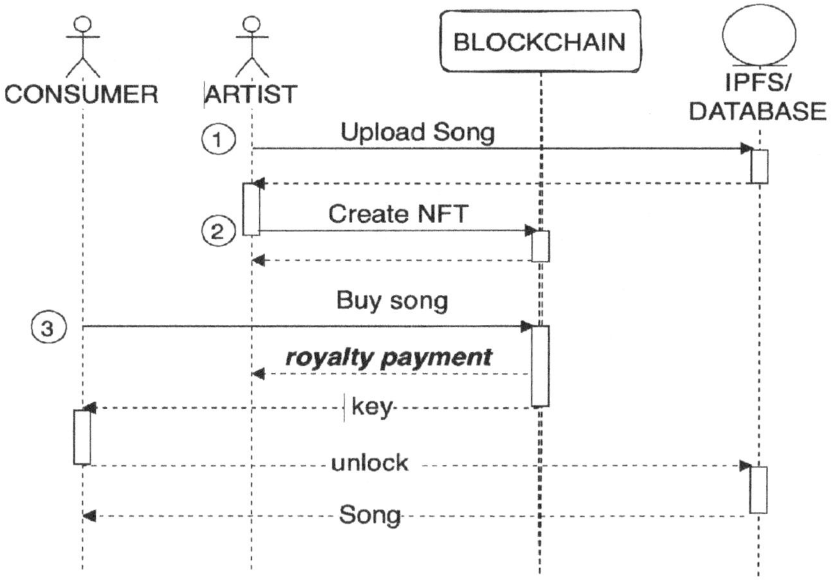

The process flow for the solution is illustrated in the sequence diagram in Figure 10, with steps enumerated below. Aria records a song and creates an NFT that owns the song's master recording. She is the original owner of the transferable NFT, which receives royalty payments whenever a fan buys a stream of the underlying song.

1. Upload Song

Aria uses the app to kick off a few steps locally. The app will encrypt and upload her new song to the Interplanetary File System (IPFS), a peer-to-peer network for storing and sharing data files. IPFS returns an IPFS file location, accessible over the internet, such as: https://ipfs.io/ipfs/Tne7fj6HDUgxv6rXqKEi-ikMK9u2NLgxxszg25pYrEKWoi.

2. Create NFT

Aria runs a smart contract by submitting a staking deposit of cryptocurrency that guarantees that she owns the copyright to the song. She also uploads the link to the IPFS file (from step 1) and meta-data, such as music genre, song length, and a standard identifier code used in the music industry known as the International Standard Recording Code (ISRC code). The smart contract contains essential functions and addresses:

> *Buy a Stream* is the smart contract to which a buyer sends payments to buy the song stream.
>
> *Royalty Recipient* is the wallet address to which royalty payments are sent when a buyer buys a song stream. These payments are temporarily locked in escrow to allow a time window for anyone to make copyright claims against Aria's song.
>
> *Offer NFT* is the smart contract for buying and selling the master recording via an NFT trade.
>
> *Copyright Claim* is the smart contract for making a copyright infringement claim against a song. The sender must submit a good-faith deposit of cryptocurrency, which would be reimbursed if the claim is deemed valid by the Oracles.
>
> *Oracles Hub* is the smart contract that tallies the Oracles' votes to reach a consensus on whether to validate copyright claims.

3. Buy Song

The third step happens when Buyer Bob sends a cryptocurrency payment to the address of the "Buy a Stream" smart contract, which triggers the following:

- Send a royalty payment to Royalty Recipient, which Aria initially owns, and,
- Send Bob an IPFS file location and private keys to decrypt the file to stream the song. The user interface software automates this process for the buyer via a play button.

In summary, Bob is buying a cryptographic key to decrypt a public file. The app decrypts the file and plays the song file.

Figure 11

ARIA SELLS HER SONG RIGHTS

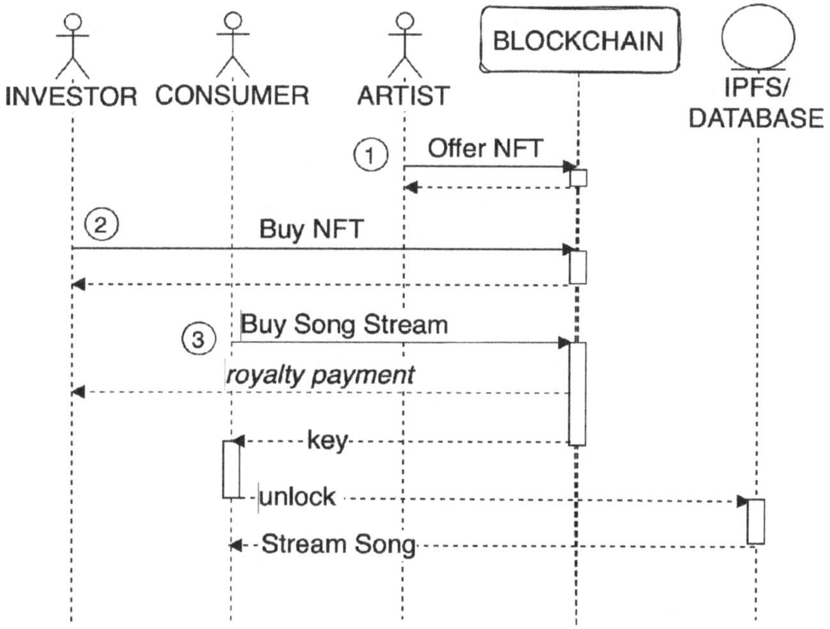

What if Aria wants to sell the rights to her master sound recording? The process involves steps shown in Figure 11.

1. Offer NFT

Aria decides on a price and sends it to the "Offer NFT" smart contract, which metaphorically puts up a "Master Recording for Sale" sign on the blockchain. Aria and potential investors all have public access to view royalty payment history on the blockchain to help arrive at a fair price. A master recordings NFT exchange is a potential blockchain business. Here, we assume a one-to-one private transaction.

2. Buy NFT

Investor Iris sends payment to the "Buy NFT" smart contract to buy the NFT. This transaction does three things: it transfers ownership of the NFT to Investor Iris from Aria; sends payment to Aria, net of transaction fees; and updates the Royalty Recipient Address to Investor Iris's address.

3. Buy Song Stream

Once the two steps above are done, Aria has sold her song rights and is entirely out of the loop. When a consumer buys a song, subsequent royalty payments will go to Investor Iris.

❦

Aria could use the same process to sell a unique song — such as a private "Happy Birthday" for a fan. Personal touches may increase brand loyalty and the chances that fans will share her music. Any time she follows someone's suggestion, that person is effectively recruited as a free marketing agent in their region because they will share the personal track with all their followers. It goes viral, where the reward system is based on a sense of belonging.

What if someone claims that Aria committed copyright infringement? There are arbitrators for cases like this: the Oracles.

Figure 12

COPYRIGHT INFRINGEMENT CLAIM

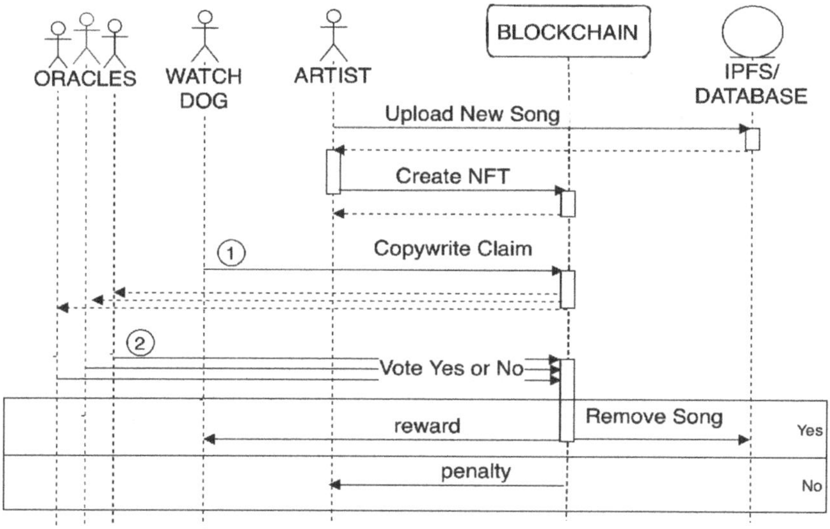

The process of handling a copyright infringement claim is shown in Figure 12. Let's assume Aria has already posted the song under review. The steps are:

1. Copyright Claim

Someone, who we'll call WATCHDOG, runs the Copyright Claim smart contract, which requires a reimbursable good-faith deposit of cryptocurrency to mitigate frivolous claims.

2. Oracles

The Oracles will consider the claim and submit their votes to the Oracles Hub smart contract, which will tally their votes to reach a consensus.

If the consensus vote is YES, the smart contract:

a. Wipes out the NFT and destroys the song's cryptographic secret keys, rendering the song file useless.

b. Penalizes Aria by taking her staking deposit, temporarily locked royalty payments, and sending both, net of fees, to the WATCHDOG.

c. Reimburses WATCHDOG's good-faith deposit.

If the consensus vote is NO, the claim is denied, deemed frivolous, and the smart contract:

a. Penalizes WATCHDOG for making a frivolous claim by taking his good deposit, net of fees.

b. Sends Aria the WATCHDOG's good-faith deposit, net of fees.

❦

A similar process with the Oracles can be used for other cases that require real-world judgment. For example, what if Aria wants to maintain veto rights over the future use of her song, such as barring her song from being played at political rallies — especially for a candidate whom she does not endorse. First, we note that an NFT contains a bundle of ownership rights. Removing rights from that bundle should lower the value of the NFT. So be it if that's the seller's choice.

One way for Aria to maintain her veto rights is via a potential "Usage Violation Claim" smart contract that grants her a way to raise claims to be voted on by the Oracles. The Oracles would then penalize the latest owner, including the possibility of stripping ownership rights.

The coding possibilities are endless. The smart contract would codify the recourse for buyers to review before buying.

Risks & Mitigation

A good plan requires anticipating risks. In this section, we consider known risks in this case study and formulate a strategy to mitigate them.

▮ Underlying Blockchain infrastructure failure.

Our solution would be for naught if one source of market power were simply replaced with another. How concentrated is the ownership or control of the underlying blockchain? Could a malicious group take over the network and thereby alter the blockchain? These questions are outside the scope of this case. Still, one rule of thumb is to use a blockchain with a high total value of outstanding currency, which makes it extremely expensive to do a 51% attack.

▮ Copyright Infringement

Decentralization is good, but some things require centralization to maintain intellectual property rights. Lawyers need someone to sue. We've shown a way to take down copyrighted content using Oracles. Individuals participating as an oracle may also be subject to legal liability. Oracles may need directors and officers (D&O) insurance. These legal matters remain uncertain.

▮ Transfer of Rights in Secondary Market Transactions

An initial offering of an NFT establishes an agreement between an artist like Aria and a primary buyer. Do all of Aria's rights transfer again if the primary buyer sells to a secondary buyer? This question falls into the legal concept of privity — a relationship between two parties recognized by law. The first party,

Aria, may or may not allow her song rights to be transferable. This is a detail for buyers to assess before buying.

▮ Bad NFT Links

Like regular internet web links that sometimes return the "404 Error: Page Not Found" webpage, an NFT that links to an off-chain database may also fail to return the expected video, audio, or image file. If the song is in a centralized database, there will be someone to hold accountable. Not so on a decentralized file storage system like IPFS. The song buyer may be subject to a loss. This risk may be mitigated by setting parameters to ensure multiple backups of files on a decentralized database, such as IPFS.

▮ File Sharing

Users might pay to download or stream a file and then share it. To mitigate such risk, digital rights management systems may use white-box or other technologies that make it extremely difficult to share usable files.

Conclusion

We've designed an approach that seeks to destroy the market power of music corporations. Of course, corporations will respond, maybe by creating blockchains of their own. One approach is to create so-called "permissioned" or private blockchains in which centralized authorities maintain control. As always, the person or entity that owns this permissioned blockchain has full control of it and potential market power. Open blockchains introduce a way to increase innovation and competition, which is ultimately a benefit to end-users. Closed or permissioned blockchains put market power back into the hands of centralized authorities. Conflating the notion of open and closed blockchains benefits those with market power.

Technologists create new possibilities for disseminating music, but it's up to creative artists to bring them to life. Music offers a cultural and emotional experience. A decentralized blockchain-based solution enables artists to share these experiences while collecting the lion's share of revenue. They also maintain control of their businesses and long-term careers by establishing a one-to-one relationship with fans.

The endgame for artists may be to work with piecemeal service providers on competitive terms. Hire a social media marketing pro. Artists might not seek to sell to a label but to retain control of their rights and build a successful long-term business on their own. With proper support, musicians can remain in charge, focusing on their art, while marketing and technology crews support creative artists by handling the infrastructure.

Case Two

Trading Bills of Lading

In a supply chain, goods move in one direction, and money moves in the other. Transacting parties don't always trust one another, and power is often unequal among them. Whichever side has stronger bargaining power can negotiate better terms, such as "buy now, pay later," to the buyer's advantage, or "cash in advance," to the seller's advantage.

Figure 13

Bills of Lading are Centuries Old

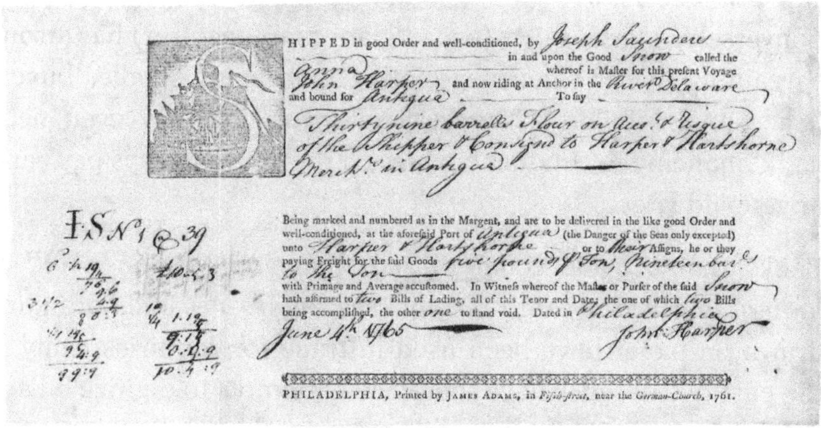

SHIPPED in good Order and well-conditioned, by Joseph Saunders in and upon the Good Snow called the Anna whereof is Master for this present Voyage John Harper and now riding at Anchor in the River Delaware and bound for Antigua To say

Thirtynine barrells Flour on Acc.t & Risque of the Shipper & Consigned to Harper & Hartshorne Merch.ts in Antigua

Being marked and numbered as in the Margent, and are to be delivered in the like good Order and well-conditioned, at the aforesaid Port of Antigua (the Danger of the Seas only excepted) unto Harper & Hartshorne or to their Assigns, he or they paying Freight for the said Goods five pounds p Ton, Nineteen bar.ls to the Ton with Primage and Average accustomed. In Witness whereof the Master or Purser of the said Snow hath affirmed to two Bills of Lading, all of this Tenor and Date; the one of which two Bills being accomplished, the other one to stand void. Dated in Philadelphia June 4th 1765

John Harper

PHILADELPHIA, Printed by JAMES ADAMS, in *Fifth-street*, near the *German-Church*, 1761.

Source: Wikimedia Commons. A Bill of Lading in 1761 Shipped in Good Order and Well-conditioned. Philadelphia: James Adams, 1761.

One-to-many or many-to-one element scenarios also create a power differential. For example, one large buyer — and no others — has strong bargaining power when negotiating with many smaller sellers. An example might be one large U.S. retailer buying hand-crafted items from many individual artisans in other countries.

Figure 14

Size Matters

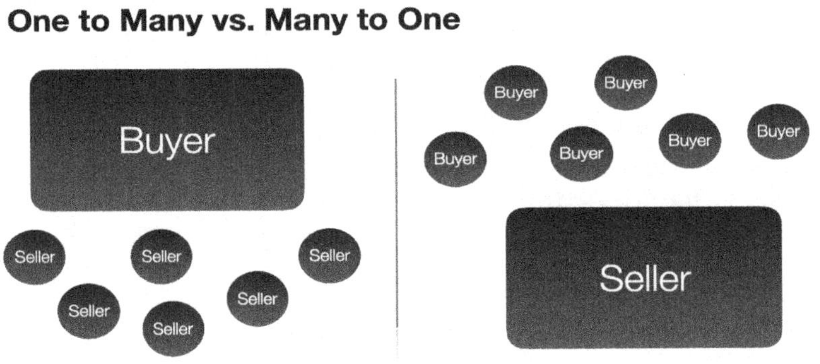

Conversely, a large seller (e.g., a giant manufacturer) has strong bargaining power when negotiating with many smaller buyers (e.g., small retailers). The less powerful side may even need bank financing to conduct business if they must always pay early or get paid later.

Banks that provide collateralized financing rely on a *bill of lading*, which is a cargo receipt (see Figure 13). Bills of lading drawn on paper have been used in trade for centuries. Only in the past two decades have there been attempts to digitize bills of lading via traditional electronic systems. These bills have yet to be standardized. Less than 1% of the world's cargo is transported with an electronic bill of lading.[1]

[1] London Shipping Law Center (LSLC) Global Webinar Series - 6/21 - Are Electronic Bills of Lading the Solution? Youtube, 5 July 2021, www.youtube.com/watch?v=j8dpDYHqvuY, Accessed 18 Apr 2024.

In July 2023, the U.K. parliament enacted the Electronic Trade Documents Act, allowing businesses that rely on English Law to exchange bills of lading and other trade documents electronically. This milestone is significant because 80-90% of all bulk shipping bills are estimated to be governed by English law[2]. Other jurisdictions worldwide may follow suit. How could blockchain solutions fit within this framework?

What if a seller could tokenize and trade a bill of lading to manage financial risk? Could blockchain technology create a globally interoperable system without a central authority domiciled in one country?

For this case study, we'll explore how a supplier, Marina, can revamp her family's export business. Her father retired and passed the company on to her. She's a technology graduate planning to exploit non-fungible tokens (NFTs), using them as collateral for bank lending needs.

Marina is eager to prove herself as a capable leader with her innovative use of blockchain technology. To do so, she plans to create a market for parties to trade goods in transit by adding sensors to her goods — the sensors are part of the Internet of Things (IoT), the collective network of connected sensors and communication devices. Sensors allow supply chain participants to monitor the location and condition of goods. This will increase their confidence in Marina's shipping business, and develop loyalty to her brand. The IoT sensors allow banks and insurance companies to monitor their collateral and financial risk in real-time. As a result, and very importantly, Marina can get cheaper financing and insurance.

[2] Mads Wacher Kjaergaard, "Boost for electronic bills of lading as UK adopts Electronic Trade Documents Act," BIMCO.org

Marina collects cash flow sooner in this new business model while creating a tradable market for NFTs. She can then reinvest in her business and enjoy a stronger position in the export trade.

We propose tokenizing bills of lading, which serve as collateral for bank lending. Then, we'll create a market to trade cargo in transit. These activities will increase the efficiency of the supply chain. Many firms have tried to develop traditional electronic bills of lading. Still, getting the entire world to use a single company's enclosed system is very challenging. A global open blockchain without centralized control may disrupt parts of supply chain finance.

Current Situation

International commerce law has codified the obligations of every party involved in the distribution of goods, from producers to shippers to intermediaries and the final buyer.

The provenance of a good refers to its origin and chronological record of its ownership, location, and other relevant information as it moves along a supply chain. Information about goods is increasingly being gathered by embedded sensors that generate location, temperature, humidity, vibrations, or other telemetry data.

There is also increasing pressure on suppliers to disclose every detail of their process, from sourcing to production, packaging, distribution, and chain of custody. From the public and policymakers alike, consumers want to know more about the carbon footprint of a good and to be able to trace every stage in the production of the food they're eating.[3] All these data may be

[3] Marshall D., et. al. "What's Your Strategy for Supply Chain Disclosure?" MIT Sloan Management Review, 14 Dec 2015.

broadcasted in real-time using communication networks such as 5G or the Long Range (LORA) network.

BILL OF LADING: IN MORE DETAIL.

The bill of lading is a transport document issued by a carrier to a shipper for the carriage of goods. It has three main aspects: it's a receipt; it's a contract; and a title document.

- Receipt

  As a receipt, it confirms that goods have been placed on board a vessel for delivery. The receipt covers the quantity and quality of the cargo.

- Contract

  This contract of carriage is between the shipper and the carrier. Many such contracts contain terms and conditions that are part of a broader overall master agreement. The bill of lading is only evidence because it's only issued when the carrier receives the goods. The master agreement to transport the goods would already have been drawn up by the time a bill of lading was signed.

- Title

  A document of title indicates that the carrier must deliver the goods to the proper or assigned person ("consignee"). The consignee can claim the goods upon their arrival at the destination port.

A bill of lading may be either negotiable or non-negotiable. A non-negotiable one identifies the name of the unique person authorized to take delivery of the goods. Conversely, a negotiable one has no specific name and is transferable. Anyone in possession of the document can claim the goods.

A negotiable bill of lading transfers ownership of goods from one party to another. It works like an open-ended contract, allowing the consigned party to transfer the title or ownership of goods to another party. This type of bill facilitates a letter of credit financing and enables the sale of goods while they are in transit. The bill of lading must be an original copy to prevent vessel operators from releasing the cargo to the wrong party.[4]

A negotiated bill of lading is used for two primary situations: the sale of goods during transit, and the sale of goods funded by a letter of credit, a document from a bank that guarantees payment for goods or services.

Success Criteria

For success, we seek the following:

- negotiable bills of lading in electronic form[5];
- tokenize bills of lading as non-fungible tokens (NFTs); and
- trade NFTs without a third-party intermediary.

Every participant in this enterprise must agree to join and play by the rules. The biggest challenge has always been that bills of lading are only legally valid in some countries. However, if one starts small, within an experimental group of industries or countries, it could undoubtedly grow and succeed.

[4] "Negotiable Bill of Lading Explained." Global Logistics Know How, 29 May 2020, maxfreights.com/negotiable-bill-of-lading-explained/.
[5] Petronilho F, Fonseca H, Zúquete A. "The State of the Art of the Electronic Bill of Lading." F1000Research. 2022;11:991. doi: doi.org/10.12688/f1000research.123856.1

Solution

For small suppliers and buyers who get poor trade financing terms, we'll build a decentralized solution so they can trade goods in transit and advance their cash flow positions.

Figure 15

A BLOCKCHAIN- AND IOT-CENTRIC SUPPLY CHAIN

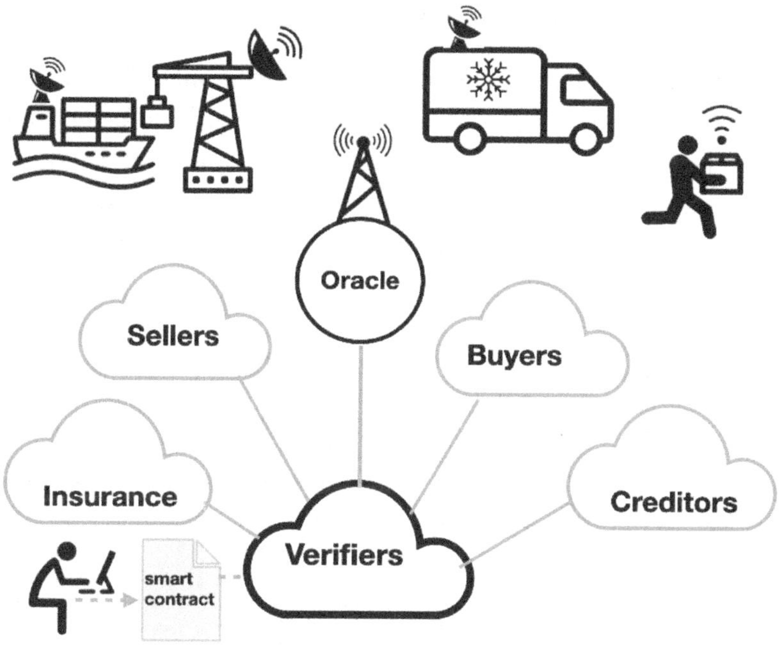

Marina's Bills of Lading

Just as with a traditional bill of lading, Marina's tokenized bills in the form of NFTs would include a cargo description and destination. There is no global standard for what information belongs on Marina's NFT bill of lading, but she can review existing documents and propose a new open standard. This would draw attention to her business — especially if the marketing department could publish her approach in a business magazine or some trade association publication.

Marina is astute and knows she can't alienate her long-standing customers and risk damaging the business by pushing them to participate in a new system. Therefore, she'll mix the old with the new. She'll scan and upload an encrypted version of physical paper bills of lading. Then, she can embed the data or image inside an NFT. Digital signatures and approvals by third-party organizations may be required.

Remember, Marina is focusing on negotiable bills of lading. She knows blockchain solutions add little value with non-negotiable bills of lading since they're not tradable. A traditional database may be more efficient when there is no need to transfer tokens without a third-party intermediary.

Figure 16

Tokenizing Bills of Ladings with NFTs

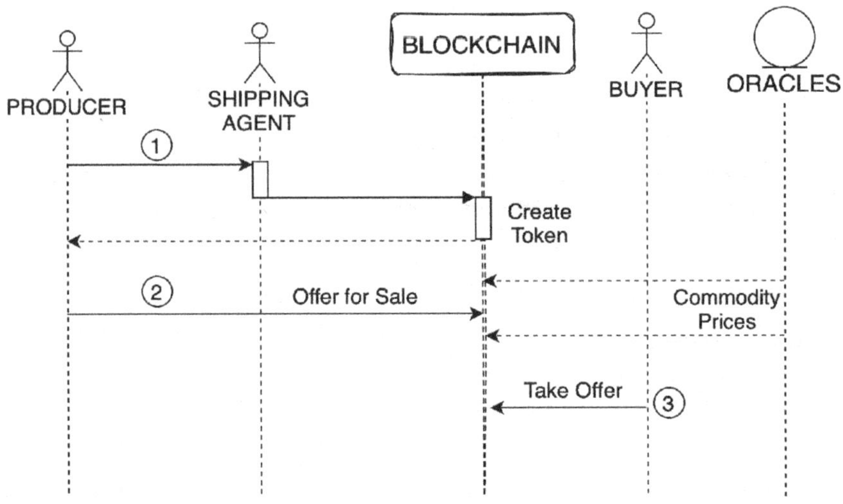

Let's see how Marina's NFT bill of lading operates (see Figure 16).

1. Create NFT.

Marina, acting as a producer, initiates the NFT by sending a message to her shipping agent to digitally co-sign a smart contract. The contract creates a tokenized bill of lading as an NFT on the blockchain. Marina, as the producer, is the initial owner of the NFT. If the NFT is to be pledged as collateral, the smart contract will allow a bank to repossess the NFT if the borrower defaults.

2. Offer NFT for Sale

Marina may sell her NFT by posting her intent on a blockchain or decentralized exchange. The price may be fixed or based on commodity prices or custom telemetry data. For example, the price offered for electronic goods containing copper components may depend on the market price of copper. The price might have risk-based adjustments if telemetry data reveal that the cargo experienced unusually high vibrations. Telemetry or IoT data has value.

3. Take Offer

When an offer is accepted, the actual price is calculated based on real-time external prices.

Risk & Mitigation

A good plan requires considering anticipated risks and formulating a strategy to mitigate them.

▮ Protection and indemnity insurance groups (i.e., "P and I clubs") don't recognize NFTs.

Electronic bill of lading documents are prerequisites for NFTs. However, electronic documents have yet to gain mass adoption. One possible way to nudge the acceptance of NFTs is to use a private blockchain — a federated or consortium blockchain — with a limited membership of mutually distrusting parties.

This type of blockchain relies on governance rules set by the group, much like a private club's house rules. Only certain members are permitted to transact in the private club. However, who would have the power to admit members? Who would have the power to verify NFT transactions? A private blockchain may eventually degrade to a traditional database application with centralized control. The question is, ultimately, who has market power?

▮ IoT devices get hacked or tampered with.

Big data algorithms need to find patterns that predict fraudulent behavior.

▮ NFT buyer fails to claim cargo.

NFT buyers are ultimately expected to take physical delivery of cargo. It's up to the NFT owner to take delivery or sell. We'd borrow from existing procedures in the futures market to handle potentially unclaimed cargo.

Case Three

Car-Sharing Service

Uber and its ride-hauling competitors are part of the new gig economy in where more and more individuals work as independent contractors rather than as full-time, permanent employees. These workers set their own hours, earning extra money on the side. Meanwhile, employers like Uber get a flexible workforce, avoid the costs of employee benefits, and circumvent labor laws that favor permanent workers. Uber created a market that matches riders and drivers to the benefit of both.

Uber collects data on thousands of rides and feeds it into their proprietary algorithms. These algorithms calculate both the rider's fare and the driver's payout. The prices reflect real-time supply and demand, and historical patterns, rather than established fixed rates. Uber has superior information.

In this model, drivers reportedly get around 50% of the fare[1]. Uber can set its own prices as an intermediary, suggesting it may have some market power. Now, suppose a driver wanted to connect directly to riders. How could blockchain technology help generate a higher payout to the driver for the same amount of work?

In this case study, we'll explore how taxi driver Travis can use a decentralized autonomous ride-sharing platform to get paid directly from the rider at the time of service and collect a larger

[1] Zhou, Yujie. "Uber Is Hiding Customer Payments from Drivers. Again." Mission Local, 16 Nov 2022, missionlocal.org/2022/11/uber-hiding-customer-payments-from-drivers/.

share of the fare. We'll disentangle Uber's business model and choose the parts that decentralization could potentially disrupt. For the sake of argument, we anticipate a hypothetical mega-merger of Uber, Lyft, and other potential rivals. Such a merger could create a firm that is more valuable than the sum of its parts by consolidating market power. Ride sharing platforms can create network effects with close to zero margin costs. A blockchain solution is a free market response to this scenario.

In this case, we are particularly interested in the data collected on every trip. We'll store this data and assign tokens to create a data exchange that rewards its creators (i.e., riders and drivers).

We propose a decentralized autonomous exchange that matches riders and drivers and runs a "frequent rider" or "frequent driver" rewards program.

Current Situation

Uber grew rapidly during the 2010s with a multi-sided business model, matching drivers and riders to offer rides in urban areas. During the fourth quarter of 2023, they booked 2.6 billion trips while collecting $19.3 billion in mobility gross booking revenue, reporting a revenue margin for mobility trips of 28.7%.[2]

As of December 2023[3], Uber is active in 72 countries and territories, with a more than $126 billion market capitalization. They disrupted taxi markets worldwide, while clashing with taxi driver unions and municipal government regulators. Time Magazine noted, "The meteoric rise of Uber, from a service in San Francisco to a billion-dollar global business, has sparked

[2] Uber. "Uber Announces Results for Fourth Quarter and Full Year 2023." Press release, 2023.
[3] "Uber Technologies Market Cap Insights." Ycharts.com.

disputes from Delhi to Rome to Manila."[4] The repercussions in the U.S. have also been significant. In New York City, the price of a transferable taxi license — known locally as "taxi medallions" — fell from $1.4 million to $80,000 between 2015 and 2021[5].

Riders register using a smartphone with the Uber mobile app. They request a ride by entering their location and destination. This information is transmitted to Uber. The driver has a driver-side app with a navigation system that guides them to the pickup and drop-off locations.

In a 2-way rating system, riders and drivers give each other ratings based on their trip experience. Riders see information about the driver and the car (i.e., license plate, name, photo, rating). Likewise, drivers may check rider ratings. The rating system increases confidence and security for both parties. Exchanging information via the app avoids misunderstandings, especially when the rider and driver speak different languages.

Uber collects numerous data points, which flow into its proprietary pricing algorithm. The algorithm calculates and then offers two prices: the fare for the rider, and the driver payout. That algorithm is a trade secret. It's reasonable to speculate that Uber's profit-maximizing strategy is to charge the highest fare riders will accept and pay drivers the least amount they will bear. Various media reports estimate that Uber's share of the fare ranges from 30% to 50%.

[4] Grillo, Ioan. "Uber Drivers Hunted down in Mexico as Taxi Unions Fight Online Competition." Time, 10 June 2015.
[5] Khafagy, Amir. "NYC Yellow Taxi Medallion Crisis, Explained." Documented, 23 Nov 2021.

Table 1.

SAMPLE OF UBER FARES AND PAYMENTS

Uber Ride Fare	Uber Driver Payout	Percentage Drivers Kept
$13.08	$ 8.13	62.20%
$20.94	$10.44	49.86%
$16.92	$ 7.72	45.63%
$14.97	$ 6.67	44.56%
$13.91	$ 6.57	47.23%
	Avg. Kept by Drivers:	49.52%

Source: Mission Local.[6]

"Uber is in an ideal position to practice what economists call first-order price discrimination – that is, charging each customer prices based on their known willingness to pay and setting each driver's pay based on their known willingness to serve,"[7] writes Forbes Contributor Len Sherman.

In terms of business strategy, Uber has an information advantage: it collects and owns data on billions of trips. Simply put, Uber owns information that others do not. New entrants may hire brilliant mathematicians to devise better pricing algorithms, but they still need proprietary data to feed into such algorithms. The algorithms are only as good as the data.

Uber's head of economic research disclosed in a Forbes interview that riders are willing to pay a higher price for a ride when their phone battery is low.[8] People with low battery levels are

[6] Zhou, Yujie. "Uber Is Hiding Customer Payments from Drivers. Again." Mission Local, 16 Nov 2022, missionlocal.org/2022/11/uber-hiding-customer-payments-from-drivers/.
[7] Sherman, Len. "Uber's CEO Hides Driver Pay Cuts to Boost Profits." Forbes, 15 Dec. 2023.
[8] Chowdhry, Amit. "Uber: Users Are More Likely to Pay Surge Pricing If Their Phone Battery Is Low." Forbes, 25 May 2016.

more likely to accept higher fares out of fear of getting stranded. However, an Uber spokesperson claims they don't charge higher fares to riders with low phone batteries. Nonetheless, if you arrive at the airport from a late-night flight with only 3% left on your phone battery, you're obviously in a weak bargaining position.

In some urban markets, riders and drivers use competing apps such as Lyft or Via to compare prices for what is essentially an identical service. Although Uber has gained a network effect, it may not necessarily have captive customers because riders and drivers can easily switch to Lyft or competitors if the price is right. Compare that to Facebook, which has both a network effect and captive customers. Facebook users won't switch to another social media platform when they've invested so much time posting photos and comments to their profiles, and most of their friends are on Facebook. On the contrary, Uber customers have nothing invested in their Uber profile and don't care if their friends use Uber. They care about price and availability. Additionally, drivers can also use both Uber and Lyft. Both riders and drivers can easily switch back and forth.

Uber has several departments with key activities. The information technology department develops and maintains the app, as expected. Operations research develops and maintains algorithms to compute fares and payouts, and match riders and drivers. Customer relations maintains and improves engagement between riders and drivers. The payments group handles payments from the rider to Uber and payouts from Uber to the driver. Human resources performs background checks on drivers and hires them.

A procurement team buys external data for the app. In 2019, Uber disclosed that it relied on Google Maps for its mapping functions and paid Google $58 million over a 3-year period.[9] A government relations department works with local laws and regulators.

Some of these activities, such as external data procurement, require someone to sign a contract.

Success Criteria

Our thesis is that a competitive ride-hauling service can be implemented and run effectively by combining elements of both real-world labor dispatching and the blockchain capabilities of a decentralized automated marketplace. Some activities, such as procuring Google map data, may need a centralized component. For now, we'll assume Google would only sell data to a real-world business, which would distribute map data to the app. Hence, we need a real-world corporation to sign contracts.

Our success criteria form a checklist as follows:

- Payments flow from riders directly to drivers,
- Trip data are the property of drivers and riders, collected and stored in a manner accessible to competing parties who pay a micro-payment to access such data, and
- A centralized organization buys map data and distributes it in a decentralized fashion.

We seek to show how a decentralized car hauling organization could function by splitting tasks into on- and off-blockchain activities.

[9] United States, Congress, House. Investigation of Competition in Digital Markets. Report No. CPRT-117HPRT47832, 117th Cong., 2nd sess., 2022.

Solution

For drivers who receive only 50% of the fare as payouts from ride-sharing companies, we'll build a decentralized solution so they can transact directly with riders and shrink the margin between fares and payouts.

We'll show how a decentralized ride-hauling organization could function by breaking down the tasks into on-blockchain and off-blockchain activities. First, we define a decentralized oracle network and then present a solution.

DECENTRALIZE ORACLE NETWORK

A decentralized oracle network (DON) is a network of inter-connected computers that run programs to collect and aggregate data from various sources. The network then provides the data to smart contracts. For example, a group of individuals might collect the daily temperature at noon in New York City's Central Park. The network then aggregates the results to prevent a single point of failure or control — to reduce the chance of error, sabotage, or data manipulation by one person. As a result, a single reliable temperature is available for smart contracts to use without reliance on a centralized data vendor.

DONs are general data services. Typically, data moves from external sources to a smart contract, but it's also possible for data to move from a smart contract to external destinations. Chainlink, the Web3 services platform, is a leading oracle firm in this space.

A hybrid organization performs both centralized and decentralized autonomous activities. We need our blockchain to communicate with the real world.

A centralized administration organization would be responsible for the functions of several departments, including governance, governmental relations, procurement, human resources, and calculation services for matching, pricing, and payouts. It would also manage decentralized payments.

Figure 17

Calling All Cars

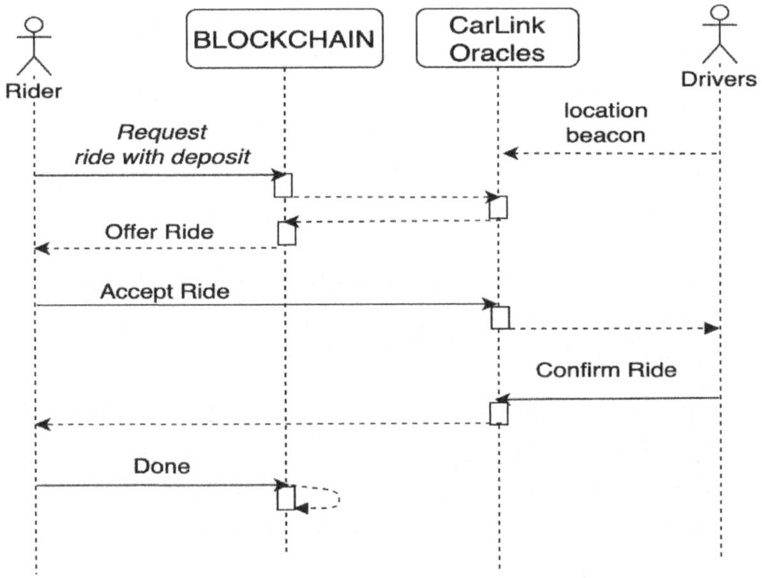

Several players will be involved in our case study. First is a decentralized oracles network, which we'll call the "Carlink Oracle Network." It will handle inbound and outbound communication with the blockchain. Next, there's a governance organization controlled by a consensus of token holders to make governance decisions. Finally, a human resources department will recruit, screen, hire, and fire drivers. Drivers must obtain a cryptographic key from the human resources department to join the network.

The process for this new system would proceed as follows (see Figure 17):

1. Location Beacon

Drivers use their mobile app to announce they're ready to pick up a rider. The app continually transmits the driver's location to the Carlink Oracle Network.

2. Request a Ride with a Deposit

Riders use their mobile app to kick off a smart contract that requests a ride. The rider must submit a good-faith refundable deposit and incur a nominal transaction fee to run a smart contract. The rider's request is routed via the Carlink Oracle Network to a network of calculation servers (Dispatchers) that compute the rider fare, driver payout, and optimal route based on mileage and an adjustment factor based on local market conditions. Surge pricing is a market reality driven by short-term supply and demand. The Carlink Oracles will return one result to the rider and driver.

3. Offer Ride

The rider is offered a fare, which they may accept or ignore.

4. Accept Ride

The rider accepts the ride and fare on their app.

5. Confirm Ride

The driver accepts the payout and confirms the rider or package pickup and drop-off locations.

6. Done

The trip is over, and the rider may tip the driver. Both sides can rate each other.

Risk & Mitigation

Let's now consider the risks and how we might mitigate them.

▮ Certain jurisdictions may block a decentralized application.

▮ Autonomous vehicles may disrupt the entire market with a startup fleet of driverless cars.

The system would need to be updated to handle an AI agent that acts as a decision-maker for each autonomous vehicle.

▮ Labor laws

If incumbent ride-sharing companies like Uber or Lyft are required to classify their drivers as employees, their costs will increase. However, it's much less likely for a decentralized app to be affected by such labor laws.

▮ Bad drivers

Bad drivers may engage in criminal or inappropriate activity that negatively affects the company. Human resources conduct background checks, but such checks are not perfect.

Case Four

Real Estate Finance

Real estate developers work on projects of many sizes and in many sectors: residential, industrial, agricultural, and commercial. These projects can take years to complete. After all the macroeconomic and financial analysis and marketing studies, the bottom line for any real estate project is how much money buyers will pay. Will buyers close the deal in a specific neighborhood for a particular building? Estimating demand — known as demand validation — is part art and part science.

In this case study, we'll focus on financing a project in a residential market. Developers, investors, bankers, and homebuyers all take risks when deciding whether to commit to a planned project. Will the developer build it, and will anyone buy the properties? Will investors recoup their money? Will a construction loan default? Will a homebuyer be the lone owner in a half-built and abandoned new development? Will local zoning laws make completing the project impossible?

The chief economist at Freddie Mac, a U.S. government-sponsored enterprise that buys mortgages, estimates the U.S. had a housing supply deficit of 3.8 million units in 2021[1].

The causes of housing deficits are many and complex. First, onerous local regulations have the power to impede construction. NIMBY (not-in-my-backyard) resistance from neighbors can also be a significant hurdle. Finally, when all conditions

[1] Khater, Sam, et al. "Housing Supply: A Growing Deficit." Freddie Mac, 7 May 2021.

seem just about right, project leaders still need all parties to cooperate and take action, even if each party is waiting for someone else to commit first.

Take the case of a group of industrial buildings being converted to residential condominiums with some commercial space. A real estate developer may plan and develop a project to the verge of being signed off. There may be a few tipping points:

▮ Missing a Lead Investor

A big-name investor who brings prestige and adds confidence to the project will inspire other investors to follow suit.

▮ Missing an Anchor Tenant

When a large, well-known company signs a long-term commercial lease, other residential or commercial tenants naturally follow. Typically, the anchor tenancy may get a good deal in return for lending their name to the project. An anchor tenant could be Whole Foods, Safeway, Cheesecake Factory, or Trader Joe's.

▮ Need More Presale Commitments

Bankers or investors may want to see at least 20% of presales in the project before committing to financing it.

COMMUNITY ACTION PROBLEM

These problems are akin to community action problems in that everyone would be better off if an action were taken, but nobody wants to make the first move.

Let's consider some questions related to this situation:

Could a blockchain application remove uncertainty and add confidence to a project?

Could investors and buyers deposit cryptocurrency into a blockchain-based escrow account?

What if the cryptocurrency is released in stages or refunded based on project milestones?

Can a group of investors take on the role of the collective lead investor?

In this case study, we'll examine how real estate developer Mason can use a decentralized autonomous system to kickstart a real estate development project by concurrently raising capital and transparently disclosing buyer commitments.

We propose overcoming the collective action problem, a social problem when it's difficult to galvanize a group to take action even though that action will benefit them. We seek to de-risk the situation — diminish individual anxiety and trepidation — and provide more transparency and trust in fundraising.

Current Situation

After five years of working for a prominent corporate real estate developer in New York City, Mason quit his job to return to his hometown of Schenectady, New York, to try his hand at leading a real estate development project. He plans to convert an old industrial building into residential condominiums, with commercial space on the first floor. His father and a few friends may provide some seed capital, but only a little. Many retired baby boomers in the area are former engineers from G.E. and IBM who, despite their age, remain tech-savvy and have kept up with the latest developments in blockchain technology. These retired engineers are a tough crowd because they don't fall for pretty sales pitches with emotional appeal. They want hard numbers.

They are not uber-rich but have a comfortable retirement, own a few rental properties, and want to invest around $100K each on several projects.

From a macroeconomic perspective, the Albany capital region of New York state isn't necessarily attracting many real estate investors. The population is proportionally shifting to sunnier parts of the country. However, real estate is local, and there are always exceptions to the rule. Good investors find these exceptions. Mason's thesis is that in the post-COVID world, people who work remotely can live away from central cities. Also, many retirees want to live in places with four seasons and no hurricanes. If Mason can demonstrate sufficient demand for this project, he can get the deal done. The proof must be compelling enough to convince skeptical bankers and investors. Mason has performed market studies showing demand for his project. He is negotiating leases with a supermarket, a national coffee chain, and a bank branch for commercial space. Signing commercial leases with high-quality anchor tenants adds prestige to the project and helps secure favorable financing.

Mason will need a variance from the local zoning board to convert the property from industrial to residential use. Mason anticipates some opposition from local residents because "Not In My Back Yard" is a factor in any real estate development. Mason cares about maintaining good relations with the community, so he has struck a deal with the city to maintain a public park across the street. The green park pleases residents and becomes a selling point for condo buyers.

It's crunch time now. The project is in a state of gridlock. Many parties have expressed interest, but their interest depends on other parties jumping in first.

Investors and lenders want to see if any condos have already been pre-sold. Condo buyers are reluctant to commit deposit money to a project with an unknown developer. Who knows? The project may never get off the ground. Potential buyers see that there are no supermarkets in the current industrial area. They are not happy about that. And the bank and supermarket will only lease if they see customer demand in that part of town.

In practice, dealmakers sometimes bend the truth to get deals done. Mason learned a few shady ways that deals would get done, such as telling buyers that a supermarket has already signed a lease, exaggerating to investors the number of presale commitments signed, or telling one bank that another bank has committed to lend.

One shady way to estimate demand is to post fake ads for condo units for sale to find out how many people will call in. The ads need to be specific enough to measure demand but not so specific that potential buyers know it's a fake ad. You also don't want them to try to find the actual (non-existent) condo building. These gambits are risky because there are few condos in the area, and it's a tight-knit community. Everyone here knows everyone else, so if Mason were caught fibbing, his reputation would be tarnished. Nevertheless, that's not Mason's style. He wants to be 100% candid about his project and won't fib to collect data.

After months of meetings, Mason has made handshake agreements with some inter-dependent conditions. A banker requires 20% of condos to be pre-sold before committing to a construction loan. A critical mass of investors would follow a "lead investor" who puts up at least $1 million. Mason estimates that getting a supermarket to sign a lease would boost presales by 10 percent. A chain supermarket will sign a lease when 50% of condo units are sold. A bank branch will open if 80% of condo units are sold or committed to be sold. The zoning board approves a zoning change on a quid-pro-quo basis if the condo association maintains the park.

One way to get this deal done may be to give everyone the misleading impression that the deal financing is nearing completion. However, Mason seeks full and open disclosure.

This illustrates the collective action problem: nobody wants to go first. Many parties are willing to follow, but no one wants to be first to invest equity, commit to buying a condo, or sign a commercial lease. Investors don't trust marketing studies, so no amount of research will convince them. Mason must prove to skeptical investors that the demand is there and ready to be tapped.

This brings us to the foreshadowed solution: a blockchain that puts trust in a transparent network.

Success Criteria

We seek to remove distrust from the marketing and capital-raising process and share real-time information. We'll provide potential investors with meaningful, transparent, and real-time metrics of project status. There are times when honesty is the best policy. Greater transparency brings greater trust and lowers risks.

Transparency is a way to overcome the lemons problem, where a good-quality used car gets lumped in with a stereotypical lot of poor-quality used cars. The owner of a good-quality used car has a vested interest in more transparency, while owners of poor-quality used cars have a vested interest in obfuscation. Mason seeks to prove to bankers and investors that he is not developing a lemon project.

The project will be deemed a success if buyers commit tokens worth 20% of the total project budget. Early buyers get a 20% discount. If the deal doesn't happen, they get their money back, net of incidental fees. The project must provide a publicly viewable number of committed unit presales. It must be metaphysically impossible for Mason to disappear with tokens. The money deposited can go only to the bank as a deposit for a construction loan or be reimbursed if the deal doesn't happen.

Solution

For real estate developers struggling to kick start a new project, we'll build a decentralized financing system so they can resolve the collective action problem by transparently validating demand while raising capital.

A blockchain solution can increase trust and transparency in the capital raising and sales process, which decreases investment risk. A blockchain solution consists of several smart contracts. First, let's list the players, then look at the smart contracts.

First, there's Mason, the Dealmaker. He controls the master contract that kicks off the process. Then there are the Buyers, the condominium buyers who make deposits towards purchases of condo units in the presale market before the conversion is completed. Next, there are the Investors who take equity positions in the project. There's also, hopefully, a Million Dollar Investor. This special participant invests one million dollars and, as a bonus, gets an equity stake in the management company that runs the condo association. Last, there's the bank or set of banks considering lending to the project. Each bank will have a public blockchain address.

Figure 18

KICKSTARTING A FUNDRAISE WITH TRUST

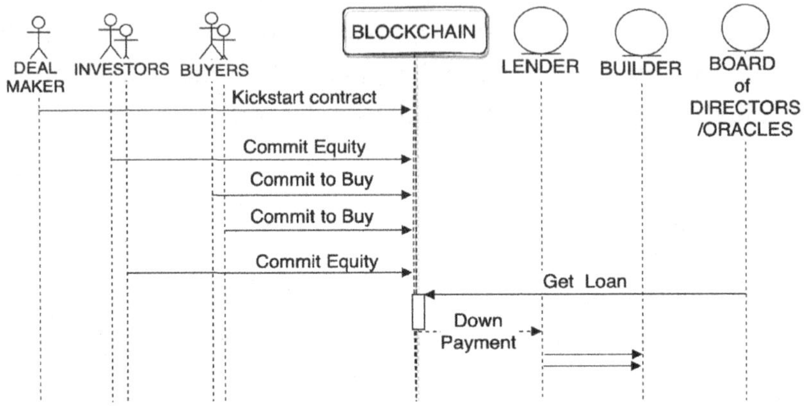

Now, let's consider the smart contracts for this case study.

▮ Kickstarter Contract

This is an administrative function that initiates the process.

▮ Commit Equity

This is a smart contract to which investors shall submit their equity commitments. Money is locked and visible to other investors. Therefore, potential equity investors may be more likely to invest if others have already done so. Transparent data on the blockchain allows all parties to assess their own risk.

▮ Million Dollar Commitment

This is a smart contract for a lead investor to deposit one million dollars in cryptocurrency equivalent value. The deposit is locked on the blockchain and visible to other investors. The lead investor gets pari-passu (a Latin term meaning "equal footing") on investment terms with all other investors. This transparent fairness encourages others to invest. However, the lead investor is granted an equity stake in the management company, allowing them to earn ongoing property management fees as a reward for being an early adopter. This setup aims to balance the interests of the lead and smaller investors, who may be reluctant to invest if the lead investor gets much better terms.

▮ Get Loan/Send Down Payment to Lender

Upon deal success and agreement with a lender, the board of Oracles accepts the loan terms and votes to release the down payment to the Lender. The lenders must be part of a pre-approved list to ensure, with cryptographic certainty, that the escrowed deposits can only be directed to a qualified bank and nowhere else. If the deal is off then deposits will be refunded to all parties. Potential bank lenders must have a public blockchain address to receive the down payment.

▮ Emergency Escape

This smart contract accepts a supermajority of votes to cancel the entire project and return all monies to buyers and investors. The intent is to have a cancel button for extraordinary conditions, such as an act of God, a natural disaster, or a war.

Risk & Mitigation

We need to evaluate risks and formulate a strategy to deal with them.

▮ The value of the cryptocurrency token falls during the project. A token could be tied to a stable value asset or a gold-backed token (see the case study on Gold-Backed Tokens).

- Extraordinary economic conditions, acts of God, or unforeseen circumstances make the project infeasible in midstream.

Under these conditions, the board of Oracles would return monies to the greatest extent possible. A supermajority of investors or buyers may cancel the project for any reason whatsoever.

Case Five

Monetizing Private Data

Surveillance Capitalism is a term coined by author and scholar Shoshanna Zuboff — it's the unilateral claiming of private human experience as free raw material that can be used as behavioral data. These data are then packaged as prediction products and sold into behavioral futures markets for businesses with a commercial interest in knowing what we will do now, soon, and later. Surveillance Capitalism operates through unprecedented asymmetries in knowledge and the power that accrues to knowledge.[1]

Many people perceive internet-based advertisements as innocuous, similar to television or radio ads, and tend to ignore them. However, unlike traditional mass-market ads, micro-targeted ads and content are fine-tuned with personalized feedback loops designed to compel each of us to take a specific action, similar to how lab rats are trained to pull levers for food rewards. Humans are trained with content that activates our primal instincts, triggering emotional responses like anger, fear, affirmation, or disgust.

Historian and professor Yuval Noah Harari suggests that as more information is collected about the world, algorithms will know individuals better than they know themselves. Businesses and governments can use this knowledge for their manipulative

[1] Zuboff, Shoshana. The Age of Surveillance Capitalism: The Fight for a Human Future at the New Frontier of Power. Public Affairs, 2019.

ends. There is a growing risk that the infomediary —YouTube, Google, Facebook/Meta — knows more about us than we do about ourselves.

The father of virtual reality, Jaron Lanier, argues that the middle class is increasingly disenfranchised from online economies. Instead of paying each individual for their contribution to the data pool, firms concentrate wealth in the hands of the few who control the data centers — and that's not the users. Lanier argues that "pervasive surveillance and constant, subtle manipulation is unethical, cruel, dangerous and inhumane." In short, this weaponized form of advertising is polarizing society, destroying democratic debate. Lanier proposes a two-way linking system that would point to the source of any piece of information, creating an economy that compensates people for the original material they post to the web[2].

Users give away private information about themselves in exchange for services they perceive to be free. As Apple CEO Tim Cook says, if you're using a product and not paying for it, "you are the product." Still, the free market has spoken: people accept the deal of paying nothing for the services they enjoy while surrendering private data.

Yet each time another bank gets hacked, we become more aware of the threats to our privacy and how our data are sold.

General Data Protection Regulation (GDPR) aims to give European Union citizens more control over their personal data. Personal data includes any information that can be used to identify an individual, such as IP or MAC addresses. It also includes combinations of features like browser characteristics, language, software version numbers, and time zones that can

[2] Thornhill, John. "Jaron Lanier on Fighting Big Tech's 'Manipulation Engine.'" Financial Times, 6 Jun. 2018.

uniquely identify a person. The critical condition is that an individual can be directly or indirectly identified. GDPR reforms are designed to reflect the world we're living in now and introduce laws around personal data, privacy, and consent. For instance, users must give explicit consent to share their data. In the US, however, Congress has tended to favor centralized industries.

Figure 19

WHO OWNS THE SURGING GLOBAL DATASPHERE (2010-2025)?

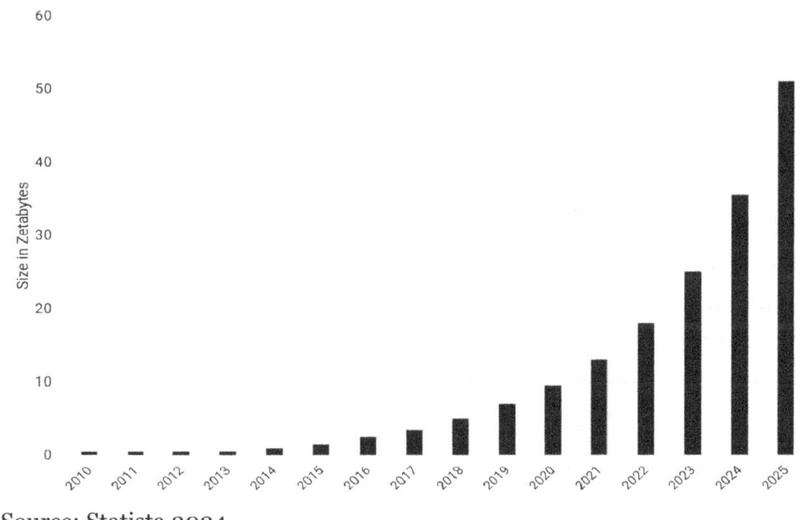

Source: Statista 2024

The rapid growth of the internet, mobile devices, internet-enabled sensors, and cloud computing have facilitated the creation and storage of vast amounts of data - the global datasphere. Nowadays, even a washing machine may have an Internet connection. Connected devices, Internet-of-things (IoT) sensors, and surveillance cameras generate more raw data. Then, raw data is processed to create even more data. Social media platforms, websites, online interactions, likes, shares, and comments contribute to the datasphere. Much of these data are owned and controlled by companies that provide free tools to users. Generative artificial intelligence (AI) model developers now scan the internet for data to train their models. If a generative AI model produces an image that looks like a Picasso painting, is that plagiarism or inspiration? Does Scarlett Johannson own rights to a computerized voice that sounds like her? Who owns private data? Some of these questions are yet to be resolved in the legal sphere.

Technological solutions often develop ahead of the government policies designed to regulate them. In the meantime, some individuals are already trying to control their private data. Business models that rely on free data could be threatened by technological solutions that block spyware or the dissemination of free browsing history.

If you know that firms are gaining market power from something they take from you for free, doesn't it make sense to stop giving it away or start charging them? What if we had complete control and ownership of our own private data? What if a blockchain token could control how your private data is used while you get paid? Rather than allowing companies to gain an advantage by using our data, why not create a system that obliges them to pay for it instead?

In this case study, we'll explore how Soledad, a private individual, can use a browser add-on app to control her private data and monetize its usage. This browser add-on will use a token to control and sell private data.

Current Situation

Soledad is a privacy advocate and the only one among her peers concerned about how companies and the government monitor online activities. She has considered deleting her social media accounts and living mostly offline without being tracked. However, she recognizes that she can't go through daily life without connecting with friends and family online, which means giving up her private data. Meanwhile, infomediaries collect personal data and sell it for profit.

Here are some examples where data about you may be used against you:

- In the US, internet service providers can sell and share personal information, such as geographic location, children's information, financial information, health information, social security numbers, web browsing history, app usage history, and communication content. Are you OK with a firm selling information about your children?

- Uber's head of economic research revealed in an interview that riders are willing to pay a higher price for a ride when their phone is almost running out of battery power.[3] Whether Uber charges higher prices is unclear, but they evidently can do so. Whether cell phone makers like Apple grant access to battery power readings via their APIs is something that could certainly change.

- Google's Nest Home Security Device was found to have a hidden microphone, which Google called an error[4]. If a company embeds hidden spyware in its products to earn extra revenue, it can gain an unfair advantage over competitors that do not use spyware. It might even be profitable to give away free products with spyware. Competitors may be forced into an ethical conundrum: choosing between embed-

[3] Chowdhry, Amit, "Uber: Users Are More Likely To Pay Surge Pricing If Their Phone Battery Is Low." Forbes, 25 May 2016

[4] Ng, Alfred, and Megan Wollerton. "Google Calls Nest's Hidden Microphone an 'Error.'" CNET, 20 Feb. 2019.

ding spyware in their products or going bust. As a result, consumers are left with no choice but to buy products with spyware that compromise their privacy.

- T-Mobile, Sprint, and AT&T are selling access to their customers' location data, reports VICE, in an investigation showing just how exposed mobile networks and the data they generate are, leaving them open to surveillance by ordinary citizens, stalkers, and criminals.[5]

Earlier, we noted that a company has market power if it can set prices above what would prevail under more competitive circumstances. We assumed that the more powerful company was a seller. However, this concept also applies to a scenario where the powerful company is a buyer. In this case, a company has market power if it can set prices below what would prevail under competitive circumstances. This implies that companies with market power are getting data for free, which they would otherwise be paying for if consumers had market power.

Success Criteria

We seek to change the power dynamic between owners and harvesters of private data. We seek an easy way for owners to collect, store, control, and monetize their private data.

Our success criteria include:

- Users control what personal data is collected, stored, and sold to or shared with various levels of anonymity
- Users have a tool to monetize their private data via tokens

[5] Cox, Joseph. "I Gave a Bounty Hunter $300. Then He Located Our Phone." *Vice*, 8 Jan. 2019

- Users may transact in micro-transactions, in units as small as one-millionth of a penny.

- Users may rent their private data to be used in accordance with their own criteria

- Users have the right to be forgotten. Users can digitally disappear if they choose to do so.

- All data is encrypted, and all searches are performed without ever decrypting the data. This means we require homomorphic encryption — an encryption technique in which operations can be performed on encrypted data without decrypting it.

Solution

For users who are concerned about privacy on the internet, we'll build a decentralized tool so they can control and monetize their own private data.

We propose a software tool to assist users in controlling their personal identifiable information (PII). The tool encrypts and stores data in an encrypted format on a server. Hence, if a cloud server is hacked, the perpetrators only see gibberish. Our solution uses Homomorphic Encryption, which we briefly describe.

HOMOMORPHIC ENCRYPTION

Homomorphic encryption is an advanced encryption technique that enables computations to be performed on encrypted data without decrypting it. The results of these computations match the results of operations performed in the traditional way. In 2009, Craig Gentry introduced the first practical homomorphic encryption scheme in his PhD thesis at Stanford.

For example, if we wanted a regular cloud server to add X+Y for us, we would encrypt X and Y, and send the result over the internet to the server. The server would need to decrypt the data to access X and Y, perform the calculation, and return a result. However, with homomorphic encryption algorithms, the server processes the encrypted data without decryption. The server then returns an encrypted result to the client, who decrypts it to view the final result. The server does computations on encrypted data without knowing the actual values of X and Y.

A practical example is in medical research. If you were asked to participate in a medical study but wished to keep your private medical records secure, you could encrypt your medical data. Researchers could then run aggregate statistics on the encrypted data without having access to your private medical data.

Figure 20

PRIVACY PRESERVING DATA SHARING

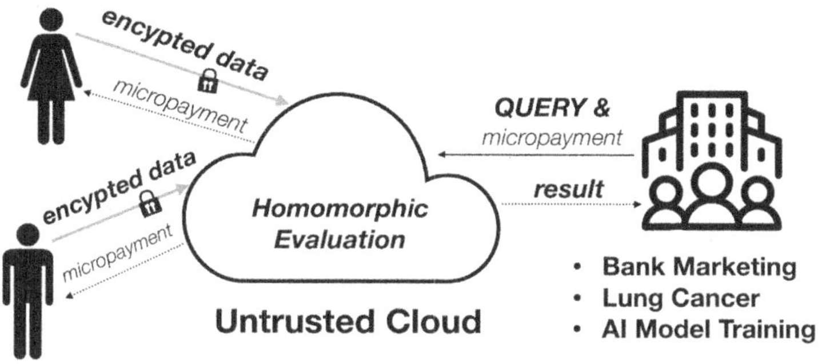

The solution entails users selecting their privacy choices, data storage, data access, and micropayments for data.

PERSONAL DATA ACQUISITION & STORAGE

Below is the process for acquiring personal data and storing it:

▮ First, users download a web browser with an add-in and built-in encryption for their electronic devices. Then, they fill in their profile and optional detailed demographic information.

▮ Next, users set privacy options to define what personal data items to disclose. They agree or don't agree to provide data elements, such as personal data from the following: smartphone and computer hardware and software, such as app usage, IP address, browsing behavior, keyboard or touchscreen entries, location tracking, access to social networks, contacts, phone charging patterns, online shopping; connected wearables, such as fitness, heart, and sleeping monitors; facial recognition metrics such as emotions analyzer, eye tracking glasses; and other home monitors include poop and bowel movement trackers, air quality monitors, and seismographs.

- A similar interface may be used for individuals to control access to their medical records, specifying what medical data elements to allow medical researchers to access.

- Last, users complete optional questionnaires or surveys on an ongoing basis, filling in as little or as much as desired. The more they provide, though, the greater the rewards.

Data is encrypted using homomorphic encryption or obfuscation techniques and stored in an IPFS database.

PERSONAL DATA ACCESS

Here's the process for how personal data is accessed in our new system:

- The company makes query requests based on profile and access to required data items.

- Consumers pre-clear permission to grant data brokers access to particular data elements or must be asked about each inquiry.

- If a data item was scanned as part of the query, then the company makes a micro-payment to the data owner.

Risk & Mitigation

Let's now consider the risks and how we might mitigate them.

- Private data owners lose their private keys - this is true of all cryptocurrencies, such as bitcoin and others. There is no good answer except to let someone hold your backup key. A bank could provide the service of safekeeping your private keys.

▮ No market develops for private data. The market for consumers to sell private data remains to be seen. One promising area for a private data management solution is a tool that lets patients control their private medical records. For example, a patient might agree to participate in a medical research project by sharing anonymized data. Healthcare data is an early use case for homomorphic encryption.

▮ Quantum computing breaks all private keys. At some point, all encrypted data may need to be re-encrypted with more advanced encryption techniques. Homomorphic encryption is expected to be quantum-safe.

▮ The cost of data storage rises faster than the value of data. The value of data needs to justify the cost of storing it. Data self-destruction occurs at a particular time, or someone buys it at a salvage value.

Case Six

Gold-Backed Tokens

The gold standard is a commitment by a country to peg the value of its currency to a fixed amount of gold. Adopted initially in England in the 18th century, the gold standard became the foundation of many international agreements. Any country that relied on the gold standard would have fixed exchange rates compared to any other gold-standard-using country.

The convertibility of currency to gold limited the power of governments to print money on a whim, which lowered inflationary fears. Yet some economists argued that these constraints also limited the government's ability to manage interest rates, lower unemployment, and respond to economic crises. The U.S., in effect, banned private ownership of gold amid the Great Depression of the late 1920s and early 1930s. England abandoned the gold standard in 1931. However, it wasn't until 1971 that the U.S. left the system's remnants, ending the convertibility of U.S. dollars to gold.

Given its colorful history, why might we want to use a gold standard today? What are its pros and cons?

In this case study, we'll explore the launch of a gold-backed crypto-token at a money management firm. Aurelio is a product manager at a money management firm working with a sizable team that includes lawyers, portfolio managers, marketing professionals, and information technology and operations teams. He is thrilled that the firm management has agreed to launch an innovative gold-backed crypto fund, which he's been internally championing. Aurelio will be the point person on a highly visible, innovative crypto fund that will deliver value to investors.

Aurelio is personally delighted about the project's social and financial impact. A gold token ultimately allows any two people, anywhere on earth, to transfer gold to each other directly without a third-party intermediary. Aurelio views the financial system as over-regulated but accepts it as part of doing business. Regulators at the Securities and Exchange Commission (SEC) can pause or halt the launch of a fund it deems non-compliant with its rules and regulations.

Laws and regulations inevitably lag technological progress. Any innovative financial product that looks different from what came before will inevitably raise red flags with regulators. Firms may spend hundreds of thousands to several million dollars in legal and administration fees to work with regulators to get a fund approved for launch. A startup firm can choose to take regulatory risks: act first and ask for permission later. However, an established firm with a good reputation and existing clients has much more to lose if it gets on the wrong side of regulators.

A gold-backed token would disrupt the multi-trillion dollar international payments business. It could unleash a wave of global economic activity. Rather than purely digital coins like bitcoin, these tokens will be backed by physical gold stored in a vault. The vault would also be publicly audited and monitored.

Let's work through an example in which Alice in Australia sends digital gold to Bob in the U.S. We'll borrow practices from the exchange-traded fund industry and explore how blockchain can make Aurelio's project come to fruition. First, let's look at Alice's current options for sending money overseas:

- Wiring money through a bank,
- Using a money transfer service such as Western Union,
- Using a cryptocurrency like bitcoin.
- Using a stablecoin like Tether, a type of cryptocurrency pegged to the U.S. dollar.

The first two traditional approaches can take several days and require the sender to pay a substantial fee. Next, using bitcoin is a fresher approach. However, Alice and Bob must accept the risk of bitcoin price volatility.

Tether (USDT) is reportedly backed by U.S. dollars in a bank account. S&P Global Ratings assesses the ability of Tether (USDT) to maintain its peg to the US dollar at 4 (constrained) on a scale of 1 to 5, where 1 is very strong and 5 is weak. "These weaknesses include limited transparency on reserve management and risk appetite, lack of a regulatory framework, no asset segregation to protect against the issuer's insolvency, and limitations to USDT's primary redeemability," reports S&P[1].

Another popular stablecoin, USD Coin (USDC), gets a better rating of 2 from S&P Global Ratings. Yet no stablecoin has achieved a rating of 1 (strongest) from S&P. Dollar-backed stablecoins have yet to be tested during a market with severe financial stress.

How about choosing gold instead of bitcoin or dollar equivalents? A gold-backed crypto token combines the old with the new.

Physical gold coins were first struck on the order of King Croesus of Lydia — an area now part of Turkey — around 550 BCE.[2]

[1] Mun, Rebecca, and Damar, Mohamed. "Stablecoin Stability Assessment, Tether(USDT)." S&P Global Ratings, Dec 12, 2023.

Some people may be hesitant about accepting bitcoin. They don't trust blockchains or any form of digital money. Others may be comfortable with blockchain technology but lack confidence in the underlying intrinsic value of bitcoin. A gold-backed token is a sweet spot for those seeking the world's most secure digital gold.

Figure 21

3 Cents for a Dollar (1913-2023)

Purchasing Power of a 1913 dollar in 2023. Source: St Louis Fed

Aurelio knows that governments around the world regulate transactions of this kind, so his challenge is to create money-like transactions without running afoul of these regulatory bodies. While it is still tricky to issue compliant gold-backed tokens, the implications for global trade and the world's marginalized, "unbanked" populations are significant. Aurelio imagines a fairer system where people in remote areas can nurture their own wealth while contributing to economic growth.

Governments have the power to regulate and ultimately seize physical gold in a vault, so a credible gold-backed token fund must operate in strict accordance with the laws and regulations of a jurisdiction. The regulatory hurdles can be onerous, so the firm that launches a fund like this would enjoy a temporary political and first-mover advantage while the fund is in its infancy. Until regulators approve any potential competitor funds, Aurelio's firm will enjoy a government-sanctioned monopoly, with time to achieve economies of scale as it grows.

Current Situation

Money is inextricably linked to the state. National currencies are controlled by central banks such as the U.S. Federal Reserve, which functions somewhat independently of the political leadership. The consensus is that political leaders are not always credible stewards for managing monetary policy. A president could manipulate the money supply for personal political gain just before an election. Henry VIII's debasement of gold coinage (where he ordered the amount of gold in coins to be lowered dramatically to fund his social life and wars in France and Scotland) led to inflation and price instability in the economy.[3]

Even if overt debasement is no longer possible, governments can still use money to exercise political power. They can pick winners and losers by manipulating interest rates, favoring lenders over savers. Politicians reluctant to raise taxes overtly might support surreptitious policies that induce excess inflation, a form of taxation through obfuscation. Why must money be connected to a particular government at all?

Government regulators closely monitor the nation's money and anything that operates like money. In the U.S., most businesses that handle money for third parties must register at the federal level as a "money service business" and satisfy an arduous regulatory approval process. Most states have their own laws and regulations for money service businesses or money trans-

[3] Downey, Catherine, and Senior, Sophister. "The Great Debasement of Henry VIII and Edward VI." 2014.

mitters operating within their jurisdictions. Governments also have the power to block others from usurping their turf. Many states in the U.S. hold monopolies on lotteries by banning others from doing the same.[4]

Governments have regulations to protect investors from fraud or to fight criminal activity. Such a rationale is more than valid. However, autocratic governments also claim the same lofty justifications for their policies. Balancing individual freedoms and state protections are ultimately political decisions. However, excessive regulations inhibit innovative solutions that would spur economic activity if money flowed more easily.

Separation of money and state may sound as strange today as the idea of "separation of church and state" to the ears of 18th-century citizens. The U.S. founding fathers saw very clearly how a church that could exercise the state's power could undermine the individual freedom of those who did not conform to the official state religion —we still see that today in some parts of the world. Similarly, a state that wields political power by controlling the money supply can undermine individual freedoms.

Success Criteria

We aim to make it so that any two parties, anywhere in the world, can instantly transfer gold ownership without a third-party intermediary. Additionally, holders of at least 100,000 tokens have the right to redeem them for physical gold on demand, subject to contractual agreements and a modest fee to defray costs. Lastly, there must be publicly verifiable proof that the amount of gold stored in a vault matches the total amount outstanding of crypto tokens.

[4] Priceonomics. "Why Does the Government Have a Monopoly on Lotteries?" Priceonomics, 17 Jan. 2014.

Solution

For anyone who likes gold, we'll launch a decentralized gold-backed token so they can transact in gold instantly with anyone in the world.

We'll borrow from best practices among traditional exchange-traded funds (ETFs) and briefly review them and how they work. First, let's discuss a traditional example of gold ownership.

> HOW ETFs WORK
>
> The SPDR Gold Shares is an exchange-traded fund (ticker symbol, GLD) representing gold bullion ownership stored in a London vault. ETFs trade as shares on a stock exchange. Investors need a stock brokerage account to buy or sell ETF shares. Unlike a regular mutual fund, which will sell shares directly to the investing public, an ETF never deals directly with the public. The investing public must buy or sell ETF shares via their brokerage account on a stock exchange. Specific institutional traders (called Authorized Participants or APs in the ETF industry) are wholesale intermediaries who move shares between an ETF and a stock exchange. ETFs transact only with APs in large wholesale sizes, typically in batches of 100,000 ETF shares, called a "Creation Unit." APs buy or sell in batches of 100,000 shares and then resell or buyback individual shares on the stock exchange. This arrangement has proven to be the most efficient way to distribute shares in the multi-trillion dollar ETF market.

A limitation of a traditional gold ETF is that an investor must use a brokerage account to buy and sell shares. ETF shares generally cannot be bought directly in fractional share amounts. Investors can't pay for something with a fractional share of a gold ETF, nor send a piece of this digital gold instantly across the globe to anyone. The stock market ecosystem is not free of third-party control like Bitcoin.

A cryptographic token represents ownership in fully audited and verifiable physical gold stored in a vault. A gold-backed crypto fund stands ready, willing, and able to barter gold for tokens, thus ensuring a steady supply of gold-back crypto tokens. The public will gain confidence in the fund if they know tokens can be redeemed for physical gold upon demand.

Figure 22

GOLD FLOW

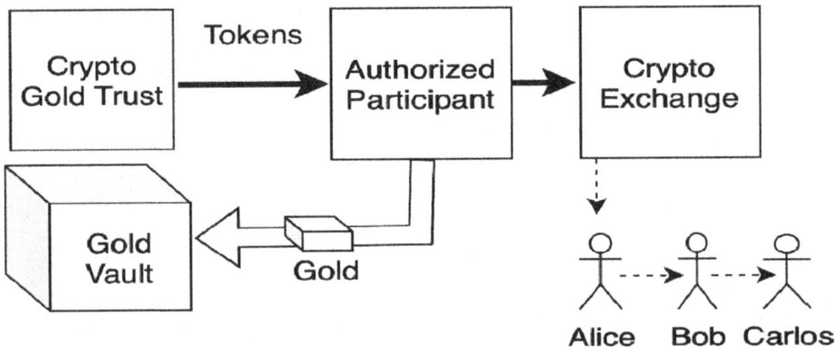

Let's look at the players in this case, which includes some special entities:

▮ Crypto Gold Trust

The Trust is the legal entity that owns gold bullion and stands ready, willing, and able to barter physical gold for lots of 100,000 gold-backed crypto-tokens ("Crypto Gold Tokens"). It has a board of directors representing the interests of token holders. The board has the power to hire and fire the fund manager, who handles the Trust's day-to-day operations.

The Trust charges a proportional share of expenses, including storage fees, plus a daily management fee. For example, if the total expense ratio is 0.25% per annum, the daily expense accrual would be about 0.25% divided by 365 days — or 0.00068493% per day. This exact method would closely follow precedents in the ETF industry.

The Trust will periodically sell a small portion of gold to pay fund expenses (e.g., the procedures of the SPDR Gold Trust). The Trust holds only London good delivery gold and maybe some incidental cash, but nothing else.

▮ The Gold Vault

This is a London Bullion Market Association (LBMA) authorized gold storage facility responsible for the safekeeping of physical gold bullion. The Trust and the Gold Vault have a checks-and-balance relationship with transparent procedures to assure the public that the number of Crypto Gold Tokens in circulation matches the appropriate weight of allocated physical gold stored in the Gold Vault.

▮ Authorized Participants ("AP")

These are institutional traders who hold a gold account at the Gold Vault and barter physical gold bullion for 100,000 Crypto Gold Tokens. No cash changes hands in these barter trades except for incidental fees.

▮ Crypto Exchange

This is a marketplace for large traders or small investors to trade whole or fractional amounts of Crypto Gold Tokens for cash or other cryptocurrencies such as bitcoin or ethereum.

▮ Alice

A small investor who has an account at the Crypto Exchange and buys Crypto Gold Tokens

▮ Bob

A coffee vendor who accepts fractional Crypto Gold Tokens directly from Alice via a decentralized blockchain transaction. No trusted third party is involved in this transaction. Bob has no account on the Crypto Exchange.

▮ Carlos

A coffee supplier who accepts Crypto Gold Tokens from Bob via a decentralized blockchain transaction.

Table 2

FUND PARAMETERS ON LAUNCH DAY

Number of Tokens in One Creation Unit	100,000
Price of Gold	$2,000/oz.
Value of One Token	$100
Value of One Creation Unit	$10,000,000
Gold Weight of One Creation Unit	5,000 oz.

The procedure for creating gold tokens borrows from the Hash Timelocked Contract (HTLC) class of blockchain conditional payments in which two parties can send conditional payments to each other with a time limit. A condition must be met for funds to be released. All tokens and gold revert to their original owners if any counterparty fails to perform their end of the bargain within a time limit. The underlying technology uses cryptographic hash functions, one of which is the SHA-256 algorithm, the details of which are beyond the scope of this book. In the following sections, we'll use the term *secret* to mean a large random number and *puzzle* to mean the "cryptographic hash" of the secret.

Figure 23

CREATING GOLD TOKENS

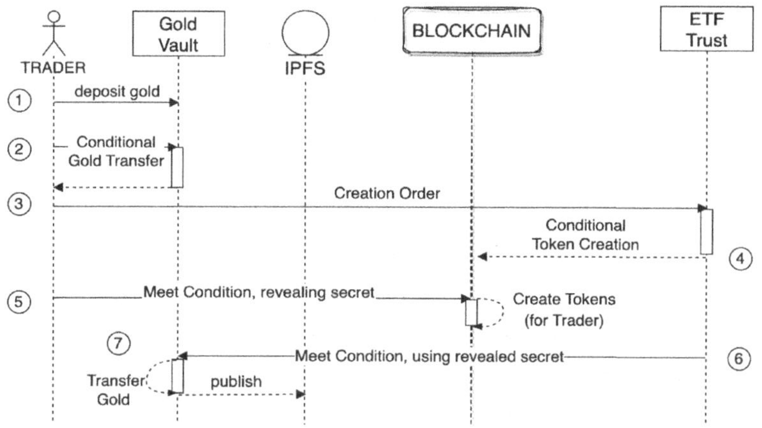

The process for creating gold tokens is illustrated in Figure 23.

1. The AP deposits physical gold bullion in the Gold Vault.

2. The AP initiates a Conditional Gold Transfer that generates a puzzle and secret known only to the AP, and authorizes the Gold Vault to transfer gold to the ETF Trust, subject to the Gold Vault receiving the secret solution to the puzzle within 24 hours.

3. The AP submits a Creation Order to request the creation of 100,000 tokens and notify the ETF Trust it will receive gold if it presents the secret to the Gold Vault within 24 hours. The puzzle is embedded in the Creation Order.

4. The ETF Trust executes a Conditional Token Creation contract that conditionally creates 100,000 tokens on behalf of the AP, subject to the AP solving the puzzle within 12 hours.

5. The AP solves the puzzle by publicly revealing the secret via a smart contract that creates and sends 100,000 tokens to the AP's blockchain address.

6. The ETF Trust reads the secret from the public blockchain and presents it to the Gold Vault. The Vault then transfers gold ownership from the AP to the ETF Trust.

7. The Gold Vault publishes the secret and the total gold it now holds on behalf of the ETF Trust.

Astute blockchain programmers will observe that the process is inspired by cross-chain token swaps using compatible Hash Time-Locked Contracts (HTLC). These smart contracts enable two parties to exchange cryptocurrencies across different blockchains without needing a trusted third party. In this case, we are setting conditions on the Gold Vault as if it were a separate blockchain from the one that holds gold tokens. However, we still need to trust the Gold Vault to perform its duties.

Figure 24

Redeeming Gold Tokens

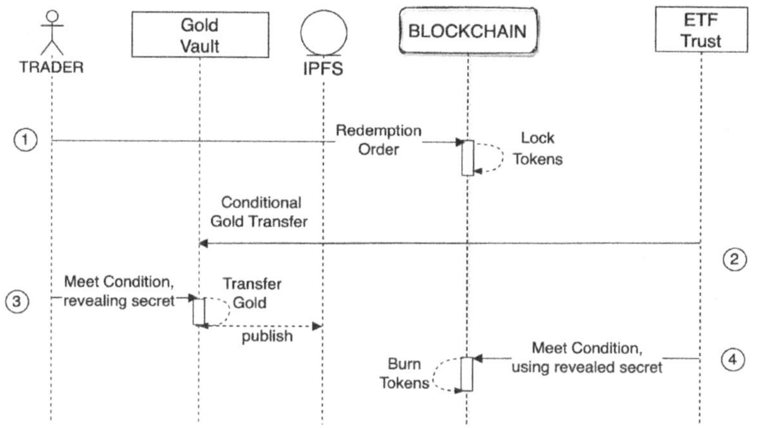

The process for redeeming gold tokens, assuming the AP has accumulated 100,000 tokens, is the inverse of creating tokens. The steps are described below (See Figure 24). The circled numbers align with the count noted at the start of each step.

1. The AP executes a Redemption Order smart contract that locks 100,000 tokens for no more than 24 hours, contains a puzzle with a secret solution known only to the AP, and destroys the 100,000 locked tokens if the secret is presented in time.

2. The ETF Trust issues a Conditional Gold Transfer to the Gold Vault, authorizing the Gold Vault to transfer gold to the AP, subject to the Gold Vault receiving the secret solution to the same puzzle within 12 hours.

3. The AP reveals the secret that solves the puzzle to the Gold Vault, which then transfers gold ownership to the AP, and

publishes the secret and the total gold it now holds on behalf of the ETF Trust.

4. The ETF Trust (or anyone) uses the publicly revealed secret to destroy or "burn" 100,000 tokens.

Anybody can audit the blockchain and the Gold Vault's publicly disclosed transaction logs to confirm that all tokens are indeed backed by gold.

FROM LAUNCH TO COFFEE

The final goal is to be able to use gold tokens to buy a cup of coffee off-chain. Below is a description of that process. It assumes transaction costs of 0.04% and that one token is backed by one-twentieth of an ounce of gold, and gold costs $2,000 per ounce on day one.

1. On day one, the AP buys 5,000 ounces of gold at $2,000 per ounce, spending $10,000,000 for gold.

2. The AP delivers gold and creates 100,000 tokens using the procedure for creating gold tokens shown in Figure 23. Because each token represents 1/20th of an ounce of gold, each token has a net asset value of $100.

3. The AP posts 100,000 tokens for sale on the Crypto Token Exchange, asking for an equivalent dollar price of $100.04 per token.

4. At some point, Alice decides she wants in on the gold tokens. She spends $100, incurring a 0.04% transaction fee on the Crypto Exchange, which sends 0.9996 token to Alice's blockchain address.

5. Alice buys a cup of coffee from Bob for $5 by sending 0.05 token to Bob via a peer-to-peer blockchain transaction. Bob uses his token to buy wholesale coffee from Carlos, who accepts Gold Crypto Tokens.

❦

We've ignored Alice's blockchain transaction fees in her peer-to-peer transfer to Bob. Ongoing development in blockchain technology, such as the Lightning Network, seeks to make these transaction fees small fractions of a penny.

Analysis

The biggest challenges of implementing a global gold coin are less technical than political. Below is an analysis of the Crypto Gold Token Case:

▮ Observation 1: The Crypto Gold Trust crosses the boundary between the virtual world and the real world.

Some cryptocurrency purists argue that a gold-back token goes against the creed of separating money from the state because the state has the ultimate power to regulate or confiscate physical gold in a vault. In contrast, it's harder to confiscate or ban strictly virtual currencies. Moreover, the real-world side of the boundary brings a host of real-world regulatory and legal challenges. New business models that use blockchains will likely perform activities on the boundary between the virtual and real worlds.

▮ Observation 2: The blockchain mining process increases computation costs to decrease social costs. The cost is justified to avoid a central authority.

Moving money around is costly. Moreover, moving fractions of a penny is only practical with solutions like the Lightning Network.

▮ Observation 3: This project requires significant off-blockchain activities.

Many blockchain business use cases must rely heavily on business domain knowledge and non-blockchain ("off-chain") activities to build an end-to-end business that uses blockchain technology.

▮ Observation 4: Crypto Gold Token holders must rely on the rule of law to enforce ownership rights.

For starters, token holders must rely on British law to enforce ownership claims to physical gold stored in a London vault. The Trust is a legal entity that must be incorporated under some jurisdiction, such as Delaware, the Cayman Islands, or Singapore.

▮ Observation 5: Crypto Gold Tokens emulate face-to-face interactions.

Throughout history, money has changed hands face-to-face without a third-party intermediary. Blockchains are seeking to take us back to those times. If it's legal to pass a piece of gold hand-to-hand, it should be legal to pass a gold token peer-to-peer. We can pass bitcoin on a peer-to-peer basis, so a cryptographic gold token is also possible. The constraint is legal, not technical.

▮ Observation 6: The total weight of gold backing up each token erodes slightly with time to pay fund expenses.

This follows the operations of State Street's SPDR Gold Shares ETF, which has an annual expense ratio of 0.40%. The Trust collects its fee by selling small amounts of gold. Therefore, one share of the gold ETF represents a slightly declining weight of gold. Gold storage is not free. Cutting corners on gold storage costs will likely increase "empty vault" risk.

▮ Observation 7: Crypto Gold Tokens partially separate money and state.

Crypto Gold Tokens loosen the link between money and the state, but the London vaults are still subject to British law. Governments worldwide might pass laws to ban their people from using gold tokens. At least we're protected from unpredictable inflation.

▮ Observation 8: Crypto Gold Tokens face extremely onerous regulatory hurdles.

The approach proposed with this Trust, which emulates ETFs, has yet to be approved in the U.S. or any jurisdiction worldwide. In the U.S., the Trust would fall under the jurisdiction of the Securities and Exchange Commission (SEC) because Crypto Gold Tokens look like securities. The Trust may fall under the jurisdiction of the Commodities and Futures Trading Commission (CFTC) because gold looks like a commodity. The Trust may fall under the jurisdiction of the U.S. Treasury's Financial Crimes Enforcement Network (FinCEN) because the Trust may meet the U.S. Treasury's definition of a "money service business." Then, one set of regulations begets more regulations.

Most U.S. states regulate money service businesses or money transmitters within their jurisdictions. The Trust may be required to register in all 50 states, an arduous process that may take 1 to 2 years to complete. Most states require money transmitters to post surety bonds. Finally, other countries around the world have their own laws and regulations.

▮ Observation 9: "Anti-money laundering" and "know your customer" rules are incredibly onerous for decentralized blockchains.

Humans have survived more than 99% of their existence without government surveillance of financial transitions. Now FinCEN "know you customer" (KYC) and "anti-money laundering" (AML) rules compel the Trust to vet anyone with whom it does business. The Trust must report suspicious activities to the government via a Suspicious Activity Report (SARs). The Trust could perform AML and KYC procedures on its directly facing counterparties (Authorized Participants) but not on downstream token owners. Regulators may require APs to perform AML/KYC on their downstream counterparties, which would be onerous.

One regulatory solution may be for APs to resell tokens only on regulatory-approved crypto exchanges, which perform AML and KYC on their customers - like trading inside a private club or a "walled garden." However, tokens become restricted within an enclosed and regulated ecosystem, which fails our success criteria, the goal of simulating the face-to-face transfer of gold coins to anyone in the world. Governments may want to keep track of who owns what to collect taxes, fight crime, or monitor the political activities of their citizens. Technological progress always escalates the arms race between criminals and law enforcement. What about privacy? What about the benefits of greater economic activity? How much freedom should we give up to Big Brother? Where is the balance? These are political, not technical questions.

▮ Observation 10. The legal domicile of the Trust and physical location of the Vault matters to token holders ("Think globally, act locally").

Token holders may be restricted from holding tokens domiciled in a particular location. A possible approach for a global product would be to issue different classes of tokens, each representing various combinations of legal domiciles and gold storage locations. Users assess political risk when they decide which class of token to own.

The legal domicile and physical location of the gold vault may include various combinations in the U.S., Switzerland, Australia, Canada, Russia, China, or India. Different classes of gold tokens would have slightly different market prices because of supply and demand and investors' aggregate political risk assessment. Considerations may include the risk that a government may tax, restrict, freeze, or confiscate gold under its jurisdiction. There is no credible way to have a gold-backed token without relying on the rule of law in one or more jurisdictions.

▮ Observation 11. Gold price volatility adds a form of currency risk.

Gold prices vs. U.S. Dollars typically fluctuate by 1% in one day and have spiked much higher at times. U.S. dollar-based merchants who accept Gold Tokens carry the risk that the price of gold may drop during the time they hold tokens. One way to mitigate this risk would be for a service provider to automatically buy/sell gold or offer a price in dollars. We can envision a very short-term futures market for merchants.

▮ Observation 12. Using a blockchain makes no sense if restrictions result in a centrally controlled system.

If a traditional database can do the job, then there's no need for a blockchain. Nowadays, plans to tokenize assets are being proposed for various assets, like stocks, bonds, real estate, cars, and music. Some of these projects are blockchains-in-name-only if they keep market power in a centralized intermediary. A traditional database could do the job. If regulators in one country ban peer-to-peer transactions, then digital gold will likely emerge in another country.

Regulatory shopping, where a Trust looks for the most suitable regulatory regime that can attract global investors, is one way to launch this fund.

Conclusion

We posit that blockchain technology has the potential to destroy political and market power held by centralized intermediaries and redistribute knowledge and power across the economy to end-users.

Many businesses may struggle to identify a use case for blockchain technology because they themselves are the intermediaries whose power is threatened by its application. Just as truck drivers may not see the value of autonomous trucks, many businesses fail to see the relevance of blockchain technology. It makes perfect sense: companies performing coordination or intermediation activities are especially threatened by a technology designed to make them obsolete. However, threats also create new opportunities.

Businesses can adapt to the evolving competitive landscape. As an infrastructure available to all, blockchain technology offers a unique opportunity for companies or entrepreneurs to reassess their strengths and weaknesses and concentrate on performing complementary activities along the boundary between on-chain and off-chain. Just as truck drivers can transition to roles in logistics in the future with autonomous vehicles, businesses can counter the impact of change by embracing blockchain technology. The key is to deflect the forces of change instead of resisting them.

Businesses that thrive in this new world will take blockchain technology as a given and devise ways to add value via complementary activities. For example, banks may not be able to charge $50 for a wire transfer in a blockchain-dominated world, but banks won't go away. They'll evolve and discover new ways to deliver value to their customers.

One crucial bank role is custody — the dull job of safekeeping your assets. When it comes to this function, dull is just fine. For example, if you lose your house keys, you don't lose your home. If you lose your car keys, you don't lose your car. But if you lose your cryptocurrency private keys, you lose your crypto money. Sometimes, it's wise to entrust our cryptocurrency to a third-party intermediary for safekeeping. A sensible practice would be storing retirement savings with a financial institution and holding liquid assets in personal wallets for day-to-day use.

One challenge of building blockchain applications is who pays for development and ongoing maintenance. A blockchain application must be part of an ecosystem with interdependent rewards and incentives for all participants.

Acknowledgments

Dmitry Koltunov's thought-provoking questions about the impact of blockchain technology on the world rather than how it works inspired me to embark on a journey to write a book I wish had already existed. Thanks to Dmitry and the wonderful people at the Startup Leadership Program.

This book would not have been possible without the support, patience, and professional advice of my editor and friend, Russell Dyer. Shula Klinger's coaching and editorial review made this book immensely more readable.

I am thankful to Professor Joseph Onochie, who guided me with an organizational structure that helped me focus on the practitioner's perspective. Michael Hillmeyer's insights on business strategy and referral to Bruce Greenwald's work were instrumental. Andrew Mazier's discussion of the intersection of power and knowledge in society influenced my views on how blockchains relate to power. Richard Anfang shared valuable insights from a senior corporate perspective.

I'm grateful to Jonathan Mohan for organizing Bitcoin meetups in New York and Solomon Lederer for conducting workshops that led to my "aha" moment about the Ethereum Virtual Machine from a computer engineering perspective. I thank J., who organized Bitdev Meetups in New York and shared deep technical dives into the Bitcoin protocol.

Finally, I want to thank the readers for their interest in this book. I sincerely hope that the ideas presented will inspire reflection, spark curiosity, contribute to meaningful discussions or debates, and inspire the development of winning business models.

Made in the USA
Monee, IL
04 November 2026

67379766R00089